# Cities and the Global Politics of the Environment

Series Editors
Michele Acuto
University College London
London, United Kingdom

Elizabeth Rapoport
Urban Land Institute
London, United Kingdom

Joana Setzer
London School of Economics and Political Science
London, United Kingdom

T0255878

**Aims of the Series**

More than half of humanity lives in cities, and by 2050 this might extend to three quarters of the world's population. Cities now have an undeniable impact on world affairs: they constitute the hinges of the global economy, global information flows, and worldwide mobility of goods and people. Yet they also represent a formidable challenge for the 21st Century. Cities are core drivers not only of this momentous urbanisation, but also have a key impact on the environment, human security and the economy. Building on the Palgrave Pivot initiative, this series aims at capturing these pivotal implications with a particular attention to the impact of cities on global environmental politics, and with a distinctive cross-disciplinary appeal that seeks to bridge urban studies, international relations, and global governance. In particular, the series explores three themes: 1) What is the impact of cities on the global politics of the environment? 2) To what extent can there be talk of an emerging 'global urban' as a set of shared characteristics that link up cities worldwide? 3) How do new modes of thinking through the global environmental influence of cities help us to open up traditional frames for urban and international research?

More information about this series at
http://www.palgrave.com/series/14897

Timothy Beatley

# Blue Biophilic Cities

Nature and Resilience Along the Urban Coast

Timothy Beatley
School of Architecture
University of Virginia
Charlottesville, Virginia, USA

Cities and the Global Politics of the Environment
ISBN 978-3-319-88518-6        ISBN 978-3-319-67955-6    (eBook)
https://doi.org/10.1007/978-3-319-67955-6

This Palgrave Macmillan imprint is published by Springer Nature
The registered company is Springer International Publishing AG
The registered company address is: Gewerbestrasse 11, 6330 Cham, Switzerland

# PREFACE AND ACKNOWLEDGEMENTS

*Blue Biophilic Cities* builds upon and extends several earlier book projects, including *Blue Urbanism* (2014), which was an initial attempt at fleshing out some of the main ideas discussed here. The chapters that follow seek to integrate more clearly the concepts of blue urbanism and biophilic cities, and also build on my earlier work on biophilic city planning and design. I argue that coastal cities offer special opportunities to foster deep connections to the wondrous marine environment around them—indeed that we must begin to understand that "nature in the city" includes those organisms, habitats and natural processes that may less obvious but are no less important or worthy of wonder. As coastal cities take steps to reconnect to the marine realm, they will have chances to build, grow and design in ways that will make them more resilient in the face of rising sea levels rand climate change.

A special impetus for this book arises from an ongoing documentary film project, which is nearing completion. With a tentative title similar to this book (*Ocean Cities*), this collaboration has led to interviews with key individuals, site visits to key blue cities, and much of the content in the pages that follow. Many thanks are due to Chuck Davis, my filmmaker-colleague, who has helped shape the ideas in the film and who has worked so creatively and diligently to make it a reality. It is hoped that this book will serve as an important supplement or companion to the film, which builds on an earlier documentary venture, *The Nature of Cities*, which aired on many Public Broadcasting System stations around the USA. We have similar high hopes for the new film. The latter relies heavily on interviews with blue–urban leaders around the country and the world, and I thank these many people for their time and for sharing their considerable insights.

A number of interviewees who shared their knowledge are due thanks. These include Josh Byrne, James Cason, Carrie Chen, Calder Deyerly, Murray Fisher, Heidi Hughes, Roland Lewis, Adam Lindquist, Theodora Long, Alan Lovewell, Jane Lubchenco, Bruce Mabry, David McGuire, Peter Malinowski, Wallace J. Nichols, Kate Orff, Bob Partrite, Orrin Pilkey, Alexander Rose, Sandra St. Hilaire, Jason Scorse, Paul Sieswerda, Peter Singer, Lindsey Stover, Stena Troyer, Harold Wanless and Julien Zaragoza. Most of the interviews were in person, often in conjunction with on-camera filming, and some were by phone.

In several places I draw from a *Planning Magazine* column I write every other month called "Ever Green." In Chap. 4, discussions of Ocearch's efforts at tagging and monitoring sharks draws on an earlier longer draft of a column in *Planning Magazine*, as does a discussion of Baltimore's Healthy Harbor Initiative in the same chapter. A great many stories and interviews are conveyed in the following pages to follow. I hope I have represented them accurately, but as usual I take full responsibility for any errors in fact or emphasis.

Some important ocean issues are not dealt with here, or only in passing. The problem of plastics and ocean garbage, and efforts to control and collect them, are not addressed, nor are the many promising efforts to generate power from the ocean. Readers specifically interested in these topics are referred to the earlier book, *Blue Urbanism*.

# CONTENTS

# LIST OF FIGURES

# CHAPTER 1

# Future Cities: The Blue and Biophilic

**Abstract** We live on the blue planet and, increasingly, an urban planet. Yet we often don't connect these realities. This chapter begins to explore these important blue–urban connections and argues that marine nature offers remarkable opportunities to promote wildness, health and healing, and a deeper sense of place. Blue cities can be *biophilic cities*. Cities around the world are beginning to seize these opportunities to overcome the "ocean blindness" that is a major obstacle to these city-ocean connections. Appreciation for nearby marine nature and wildness will continue to grow, and planning and design will increasingly reflect this.

A recent story in *The Economist* declares in its opening sentence: "Earth poorly named."[1] It ought to be called the water planet, or just blue. More than two-thirds of the surface of the planet—a vast area of the planet's living space—is water, so we have named it poorly indeed. Much of our experience as a species has been in the terretrial realm, where we have lived and evolved, so perhaps it is understandable that we have taken a blinds eye. Our expectation of the oceans has been for them to perpetually offer up a bounty of seafood, to serve as receptacles for human pollutants of many kinds, and act as transport highways, connecting cities and settlements throughout the world, linking cultures and economies.

T. Beatley, *Blue Biophilic Cities*, Cities and the Global Politics of the Environment, https://doi.org/10.1007/978-3-319-67955-6_1

These same qualities of invisibility, or "ocean blindness" as *The Economist* calls it, are what largely explain why oceans today are in such trouble. We have difficulty seeing the ramifications of the many ocean degradations we have wrought and their long term cumulative effects (though this is becoming increasingly evident). Out of sight, out of mind—though marine health and marine life may be nearer to urban residents than we think (and potentially much more "seeable" and "knowable," this book will argue).

Just as we are the blue planet so too are we increasingly the urban planet. As a city planner this does my heart good, and while cities today face a range of problems and challenges, many believe they represent the best hope for improving the quality of human life and expanding economic and social opportunity. As we seek to create a more sustainable world, cities must increasingly be a key element, perhaps the key element in such strategies. We argue increasingly for the need for denser, more compact cities, where it is possible to walk everywhere, reducing dependence on automobiles. We seek to develop more sustainable ways to feed and house urban populations and we invite more sustainable forms of infrastructure, including bikes paths, public transport and renewable energy.

The goal of creating more dense, compact cities in turn raises a question of the role of nature in urban life. Can we design and build more sustainable places in ways that maintain, indeed foster, new connections with the natural world. At the core of this goal is a belief (a key one that I hold, which will be explored in the pages that follow) that we need nature in our lives. We need to have daily, indeed hourly contact with nature. Nature is not something optional, but absolutely essential to leading happy, healthy and meaningful lives.

At an intuitive level many of us appreciate the positive benefits of nature. A stroll in the woods, or time spent tending a garden deliver clear and visceral improvements in mood, in reducing stress and in inducing more creativity. Evidence has been mounting showing that contact with nature has substantial mental health benefits and is a potent antidote to chronic stress.[2] We also know there are many other benefits—evidence from psychology suggests that we are more likely to be generous in the presence of nature, more likely to cooperate and more likely to think longer term.

Rachel Carson wrote eloquently about the importance of wonder in our lives, and here nature is uniquely suited to it. In her influential essay (later published as a book) she hoped for every child to have an indestructible sense of wonder "as an unfailing antidote against the boredom and disenchantments of later years, the sterile preoccupation with things that are artificial, the alienation from the sources of our strength."[3] For Carson, the marine edge was an especially potent and powerful place for stimulating this wonder, and for sharing it with others, in beautiful books such as *The Sea Around Us*.[4]

In the terrestrial realm we are privileged that enjoying nature can take many forms, such as watching birds and listening to their songs, walking in an urban forest, planting and tending a garden, perhaps in one's front or back yard, or balcony if one lives in a high-rise building. But there are also many forms of marine nature nearby that we must begin to better recognize as opportunities for biophilic connection and that is one of the key messages here. Much of marine nature is mysterious and difficult to see because it is inaccessible, under water, far away. But in coastal cities around the USA and the world, New York to San Francisco, from Seattle to Rotterdam, from Singapore to Perth, urban populations have remarkable amounts of blue nature, remarkable amounts of blue wildness nearby. Such marine nature must be seen through the lens of *biophilia*, and cities on the coastal or marine edge understood as *blue biophilic cities*.

E.O. Wilson, Harvard University biologist, entomologist and conservationist, deserves much of the credit for introducing the concept of biophilia, first to the environmental and conservation community, then more generally to broader society. Biophilia refers essentially to the innate connection we have with nature, our innate affiliation and emotional bonds with the natural world. We have co-evolved with nature, biophilia says, so it is not at all surprising that we tend to be more at ease, happier and more creative when we are surrounded by nature.

What are Biophilic Cities? They are cities that are nature-rich or nature-abundant, of course—cities with extensive numbers of trees and amounts of greenery, where wildlife is welcomed in, where neighborhoods make it easy to spend time outside. These are cities frequently described with reference to their natural qualities—cities that have achieved a high percentage of tree canopy cover, for instance, or a high percentage of residents living in close proximity to a park or greenspace.

In 2013 we launched a new global Biophilic Cities Network in an effort to connect cites putting nature at the center of their design and planning, sharing stories, comparing notes about effective tools and helping to inspire each other and to advance the global biophilic cities movement (Figs. 1.1 and 1.2).[5]

**Figs. 1.1 and 1.2**   Singapore has emerged as an exemplar of a biophilic city and has changed its official motto from Singapore Garden City, to Singapore *City in a Garden*. Image credits: Tim Beatley

A number of cities in the network, notably Singapore and Wellington, have done much to study, raise awareness about, and conserve and protect marine environments around them. They are inspiring stories, to be sure, but they are only emerging examples of Blue Biophilic Cities. There is much more to do and many more ways coastal cities especially can connect with and support the marine realm.

## THE BLUE CITY IS A BIOPHILIC CITY

How does, or how could, biophilia manifest in a marine or coastal urban context? Blue Biophilic Cities can be described as cities that do not ignore a marine context, but rather celebrate it. They are cities that seek to appreciate marine nature and understand that much of the biodiversity and nature around will be marine.

More specifically, blue cities that are Biophilic Cities—Blue Biophilic Cities—could be described as follows:

- cities where residents actively spend time on, near, or underneath the water near and around them, whether boating, sailing, strolling or hiking near the water's edge, participating in a variety of recreational pursuits and hobbies, from scuba diving and snorkeling to ocean and harbor swimming that bring them in close contact with the marine world and help to shape a sense of connection to the marine environment;
- cities that seek to maximize moments of awe and wonder and understand the marine world as a nearby place of immense biodiversity and majesty, recognizing it as a source of opportunities for urban experiences that soften, uplift and deliver joy;
- cities that view nature and ecology holistically, that put marine life at the core of their view of the natural world in which they sit;
- cities that actively connect residents to the marine world and that embrace a sense of caring for and protecting marine life;
- cities that attempt to reduce individual and collective consumption that negatively impacts the marine world, both near and far. Blue Biophilic Cities understand they are duty-bound to take steps to reduce the size of their ecological footprint on the world's oceans. There is an ethical duty to reduce the ecological impact but also an affirmative responsibility to do what is possible to actively conserve and protect the marine world. Blue Biophilic Cities establish marine

protected areas where it is possible locally and do what they can to assume active leadership and support for larger reserves, including those potentially hundreds or even thousands of miles away;
- cities that work to include marine education in schools and that seek to educate all citizens about marine life and the threats and pressures currently being experienced. Achieving a basic (and even advanced) level of marine literacy is an important goal in all Blue Biophilic Cities.

It might also be said that these cities seek to shape a connection to marine nature that is "whole of life." That is, a connection that begins at an early age, runs through childhood and extends into adulthood and into one's senior years. It is a love of the marine realm that is long-lasting, deep and fairly continuous. Richard Louv has written compellingly about the ways that children today grow up disconnected from nature—time spent outside in nature replaced with screen time, an inability to recognize common species of flora and fauna, parental worries about safety and a shift away from teaching natural history, among the many causes.[6] In Biophilic Cities exposure to and learning about nature starts at an early age and continues throughout life. Schools include hands-on outdoor learning and integrate nature into their curricula at every stage. And these opportunities continue in later years as we increasingly recognize the benefits and meaning nature takes on. Opportunities to engage in blue nature, it is argued here, ought similarly to be available throughout one's life.

It is important to recognize the role of the blue in the biophilic, and that is a major theme of this book. We are the blue planet, as Sylvia Earle and others frequently tell us. And yet we seem to need constant reminding of this. Cities, especially it seems, have little explicit recognition of their marine position in the world, even when they are perched on the edge of sea.

As former National Oceanic and Atmospheric Administration (NOAA) administrator Jane Lubchenco said recently in an interview for our film about blue cities, "Oceans are essential to all life on Earth."[7] Yet they are in serious trouble and in quick decline in many ways and in many dimensions. Oceans are "warmer, more acidic, hold less oxygen, [are] more impoverished than ever," and the biggest problem Lubchenco argues—bigger than climate change and overfishing—is the lack of understanding, the lack of public awareness about oceans and how essential they are to our collective future. Half the world's oxygen, she says, derives from oceans, and some 3 billion people on the planet rely on the bounty of the sea for their primary source of protein.

There is little doubt that ocean and marine environments are bearing much of the brunt of population and development pressures. Contaminated with immense volumes of plastic and other waste, and absorbing much of the world's carbon, it is showing many signs profound degradation and decline—coral reefs dying in the face of acidification, fish and marine organisms shifting in the face of rising sea temperatures.

July 2017 witnessed a dramatic event as an iceberg "the size of Delaware" broke away from the Larsen C ice shelf in Antarctica.[8] A 2016 study published in *Nature* sheds some light on the significance of melting Antarctic ice. Modeling earlier climate eras, the ominous conclusion is that a melting Antarctica gets much of the blame for those historic high sea levels: "Antarctica has the potential to contribute more than a meter of sea level rise by 2100 and more than 15 meters by 2500, if emissions continue unabated."[9] Even a meter of sea level rise will be too much for cities such as New York and Miami to handle (Fig. 1.3).

On the other hand, there are immense benefits and ecological services we gain from oceans, though many may be in decline or diminished in the future. The blue economy is worth about $21 trillion in annual income, according to the UN.[10] We can't afford to ignore the ocean, ecologically or economically.

## BLUE WONDER, MARINE MAGIC

We need healthy oceans and I believe most people would care about them if given the chance and if fully aware of their plight, what we are losing and the impact their destruction is having on us. Why this lack of awareness? Some of it, perhaps most of it, derives from the physical and emotional disconnect from the watery, marine realm. It may not be far away in terms of distance, but worlds away in terms of perception and emotional proximity. "Out of sight, out of mind," explains a lot says Lubchenco.

Filming recently on the waterfront in Vancouver, we heard from Christianne Wilhelmsen, director of the Georgia Strait Alliance, about the challenges of fostering ocean awareness even in a city where water is such a given. "Look at any brochure of Vancouver, it will show water, guaranteed. Yet we still exist in a highly urbanized environment where people can grow up never really recognizing that they are part of an ocean." And as Wilhelmsen says Vancouverites don't necessarily see how an ocean context

**Fig. 1.3** Blue biophilic cities seek to rethink the many ways they impact or connect with the marine world, including through the harvesting and consumption of seafood. Image credit: Tim Beatley

affects them directly, and they don't appreciate the many ways that their actions and behaviors affect the condition of that ocean and the organisms living in it. These are perennial challenges it seems.

The good news is that there is an urban marine renaissance under way, a rediscovery of the urban marine setting, and that many cities around the world are recognizing the importance of their water and ocean settings and taking impressive steps to connect citizens to them.

Much of this work is happening through local governments, but a lot of it is also happening through the remarkable work of nonprofits and other local harbor and ocean conservation groups. These include the work of the Georgia Strait Alliance in Vancouver, and Friends of the Seattle Waterfront. They include the work of the Healthy Harbor Initiative in Baltimore and the Waterfront Alliance in New York City, itself a coalition of 950 organizations. And there are a variety of new and creative initiatives under way to connect residents, to celebrate marine nature, as well as to grow and build in ways that moderate future risks.

But there are immense obstacles of course. "Ocean blindness" characterizes much of our official urban and environmental planning. Maps of parks, green spaces and biodiversity rarely extend beyond the shore's edge and we carry with us, unfortunately, our mental maps that are highly terrestrial in their bias. In an interview with Murray Fisher, founder of the New York Harbor School, Fisher spoke of these mapping and perceptual biases. "When people are doing work about nature in New York City they often forget about the harbor. We're just trying to get that back at the table, we want to have the harbor back at the table when discussing planning. And we'd love those maps [of urban nature] to go underwater and to show what's there."[11] We have a poor understanding of baseline marine nature and biodiversity in cities such as New York. A blue biophilic focus might (we hope) overcome these realities.

A major theme of this book is that the two realities—of a wondrous marine nature, a therapeutic blue world that enhances the quality of urban life on the one hand, and the increasing dangers of sea level rise and climate change—are indeed reconcilable. They are both real and important streams of reality that coastal and marine cities must confront, and can. Often the solutions (many I outline in this book) can respond to and address these two realities at once—we can prepare and adapt to sea level rise and the dangers of coastal living precisely through design and planning and other initiatives that bring us closer to the wonderful nature around us.

Cities, then, can and must aspire to a new more integrated vision of becoming both blue and biophilic. Cities of the sea, as I am calling them here, compellingly merge the principles of blue urbanisms and biophilic cities.

## WHY AS URBANISTS WE NEED AND MUST CARE FOR OCEANS

That water is an immense asset for blue cities is undeniable. And evidence is emerging about how deep and profoundly we are drawn to being around water. Wallace J. Nichols, author of bestselling book *Blue Mind* has been a pioneer in pulling together the evidence we have about the therapeutic and health benefits of water.[12] *Blue Mind* is the first book of its kind making the case for contact with water in all of its different forms and formats, from swimming pools to sensory float tanks to beaches and shorelines.

Few people have done more to emphasize the benefits of water than Nichols and he convenes a conference every year that pulls together researchers and activists with an interest in the power of water. *Blue Mind* is a remarkable book that summarizes the literature and science surrounding how we experience and react to water in all its forms and places. It gives a compelling case for beginning urban planning and design from a water's vantage point—oriented parks, public spaces, viewsheds, placement of streets and buildings—in ways that maximize access and connection to water. Many cities are taking this philosophy and approach (Fig. 1.4).

A sea turtle researcher and conservation advocate, J lives with his family a few feet from the beach in Carmel-by-the-Sea, south of San Francisco. We filmed J on that incredibly beautiful stretch of coastline, a place many are drawn to walk and stroll. J speaks a lot of the contrast between what he calls "red mind" and "blue mind." Red mind refers to the too-oft state of mind where we are harried and busy, worried, stressed out, struggling to meet deadlines, scanning the multiple electronic screens we most likely possess.

Our blue mind is the opposite—a mental state where we are relaxed, where we are replenished and restored. "We get this restorative quality," Wallace tells me, "from many different things. It could be a great meal, it could be talking with our friends, or it's art, or music." But there is something especially beneficial about water. "Water seems to hold a special place."

Nichols go on to explain the evolutionary basis for the restorative qualities of water. Water has been an "imperative" for the human species, something essential for survival. "So when we're by the water that itch is being scratched, that need to be positioning ourselves relative to water." Even

**Fig. 1.4**  The sea lions on Pier 39 in San Francisco are a major attraction, and a glimpse of blue wildness nearby. Image Credit: Tim Beatley

more profoundly, Nichols understands water and water-rich settings are the "backdrops for our lives":

> It gives us a backdrop to romance and big ideas. It allows us to hear our own thoughts more clearly. It's a source of solace, peace, a sense of freedom, a place to grieve, to mourn...So it's a backdrop to every great part of our lives.

Water, waves, beaches, harbors, rivers, all of those watery realms of life that play such important roles are places we are hardwired to want to be and to enjoy. "Water gives life, it makes life possible, but it also makes life worth living," Nichols says.

The academic research supports what we feel when we are near to water. There is evidence from large panel studies that show we report higher levels of health and wellbeing the closer we live to coastlines, controlling for other variables.[13] These effects seem especially apparent when we are living within five kilometers of a shore's edge. And at least some of this improved health seems to be the result of coastal residents getting greater levels of physical exercise.[14] Many of the Blue Biophilic Cities described herein reflect commitments to providing physical access and opportunities for time spent strolling, walking, hiking, running, shell-collecting, or just reflecting and would confirm these research findings. There is a kind of blue medicine and blue therapy in the waters and shore edges around blue cities.

## MEDICINE AND WILDNESS

Few stories are as convincing about blue therapy as some of those coming out of the world of surfing. Nichols writes about the former army veteran Bobby Lane who returned from war with post traumatic stress disorder (PTSD) and an inability even to sleep. Suicidal, Lane participated in a surf therapy program that changed his life and literally, he says, saved his life. I heard Lane describe first hand these gut wrenching experiences at the Blue Mind conference in Washington in 2015.

Operation Surf is the program Lane participated in, and is a terrific example of the power of water. Aimed at veterans experiencing PTSD and depression, and often with serious injuries, the program teaches participants to surf by pairing them with experienced surfers. The program takes place over six days and is run by the nonprofit Amazing Surf Adventures (ASA). At least one study shows that this form of "ocean therapy" leads to a significant reduction in depression and in PTSD symptoms, even 30 days later (though the reduction in symptoms lessens).[15] This surf therapy also leads to a significant increase in a sense of self-efficacy. A wonderful short film (made by Kellen Keene) presented on the ASA website provides first hand testimonials of the life-altering effects of Operation Surf. One veteran speaks of how surfing changed his perception of his injury, while another speaks of the smiles that spread over him. Lane speaks of the "sense of peace" that he experienced and how he no longer had the desire to commit suicide.[16]

Efforts in cities like Boston and Seattle to re-establish physical connections with water in part reflect this view of the profound health benefits that contact with water provides. In Boston, the example of Spaulding Rehabilitation Hospital is telling. Relocated from an older site, the new facility seeks to take full advantage of its harborfront location. It faces the water, with patient rooms maximizing views of the water and providing a physical connection to Boston's HarborWalk. The profound healing powers of the harbor are not lost on doctors, nurses and the designers of the next generation of hospitals and medical facilities.

The blue provides elements of wildness and adventure that are hard to find elsewhere. And wild spaces and places that are closest to where urban residents live are often in the marine realm. Murray Fisher, founder of the New York Harbor School, speaks passionately of the prospect of re-wilding marine environments, including New York Harbor. One of his signature initiatives, the Billion Oyster Project, has this as a goal.

Marine environments, near and far, though with a history of degradation, remain remarkable places of wildness. As municipal sewage is better controlled in cities like New York and Boston, and as water quality improves, they become yet wilder. Fisher describes the re-wilding New York Harbor, as a "powerful concept," one that can bring inspire and motivate and bring people together. And where else will re-wilding in cities be feasible? Harbors and near-to-city waters represent "one of the only places where you could re-wild a natural area without disturbing the existing human uses."[17]

Christianne Wilhelmsen speaks of the importance of this blue nature to the health and quality of life in cities. Orca whales are residents of the waters around Vancouver and are the logo for the Georgia Strait Alliance. When the orcas appear, the "entire city shuts down and people run to the shore." It is clear that when given the chance urbanites respond to marine nature in a deep emotional way.

Similarly the residents of New York City are re-cultivating a love of whales, in this case humpback whales have been returning to the waters of this city, and for a couple of years in a row, the city's harbor. Summer of

2017 saw humpbacks return to the Golden Gate Bridge in San Francisco, offering similar delight to residents there.

We now have evidence about our innate attraction to nature, our desire and need to affiliate with nature and living systems, the core of the idea of biophilia. Harvard biologist E.O. Wilson has done much to make this idea mainstream, and much research has been done lately in psychology, medicine and public health, even economics, to bolster the evidence.[18]

The reactions of New Yorkers or San Franciscans seeing humpback whales, or Vancouverites seeing orcas, are visceral demonstrations of the power of nature, and the innate pull it has on us. We don't often talk about wonder and awe in urban planning circles, but we should. The marine life all around us can serve as a constant source if we let it and if we actively work to cultivate and foster a sense of awareness of, and care about, nature.

It is telling that some of the biggest draws in coastal cities involve some connection to marine nature. The California sea lions on Pier 39, in San Francisco, draw hundreds of visitors every day, for instance. Originally they showed up unexpectedly, lounging on the boat docks of the marina. Being heavy animals, the pier now has a special set of reinforced floating docks to accommodate them, recognizing their permanent status as occupants of the San Francisco waterfront.

An equally important dimension of a blue biophilic city is a more ethical one. While we can and must emphasize the mental and physical health benefits of a proximity to water, and the many other benefits we gain from the oceans, we must be careful not to leave our argument there. It is a key premise of this book that ocean nature is a significant ingredient in creating a meaningful urban life. Yet, we hope that by fostering and strengthening these connections with the marine world we will begin to see and understand an ethical obligation to protect it, to conserve it and to exercise restraint on how we use and treat the marine habitats around us. There is a biocentric ethical sensibility—a recognition of the intrinsic moral worth of marine nature that must be given importance alongside the many more anthropocentric reasons. This larger urban–ocean ethic can in turn serve as the basis for supporting (and advocating for) ocean species, habitats that may exist many hundreds or thousands of miles away and for which direct personal connections may be harder to see. I have much hope that in fostering new connections to the nearby sea, a larger, ethic and ethos will emerge (Fig. 1.5).

**Fig. 1.5** A goal of Blue Biophilic Cities is to provide extensive contact and connection with the marine world. Here joggers in Coogee Beach, near Sydney, Australia, enjoy a remarkable coastal walkway with dramatic views. Image credit: Tim Beatley

Many cities around the world are re-discovering their harbor and coastal settings and investing in them. There is a growing re-assessment of marine nature as an incredible resource to fully take advantage of the opportunity for urban residents to connect to wild nature and the opportunity for awe and wonder in daily urban living. Investments in blue–urban nature pays immense dividends, in terms of quality of life, opportunities for recreation, and mental health benefits. The water around us in cities offers remarkable respite and therapy, even a passing glimpse of water provides important benefits.

That the planet is grappling with serious environmental and social challenges goes without saying, and the future vision and agenda of blue biophilic cities will fall within this larger global frame and challenge. In 2015, countries adopted the Agenda for Sustainable Development, and 17 Sustainable Development Goals.[19] Several of these goals are of special relevance given the topic of this book, including Goal 11 (Sustainable Cities and Communities) and Goal 14 (Ocean Conservation). There is every hope and expectation that the specific ideas contained in this book—and the vision of cities more connected with, and working to actively conserve and cherish the marine nature around them, will help in many ways to advance these goals.

## SOME CONCLUDING THOUGHTS

The concept of biophilia holds that as a species we have co-evolved with nature and that we carry with us our ancient brains that reflect this deep evolutionary history. We are happier, healthier, able to lead more meaningful lives when in close contact with nature. The evidence mounts that we are even likely to be better human beings when we have nature around us: we are more likely to be generous, to be cooperative, to think longer term when we are in the presence of nature. The nature we need must be all around us where we live and work, not only in a distant site we visit once or twice a year. We need daily, even hourly doses of nature. The insight of biophilia has ignited a global urban movement called biophilic cities, the vision is one of compact, dense cities that are sustainable and resilient, but immersed in nature. Biophilic cities put nature at the center of design and planning and take a whole-of-city approach that recognizes the necessity of nature at all urban scales—from room or rooftop to region and all the spaces in between.

The marine realm represents a complex, wondrous place of nature near where millions of urban residents live and work. There is great potential for this blue nature to shape and influence in many positive ways the quality of future urban life. We know, and there is increasing supporting evidence of, the many ways in which the urban blue is biophilic. Blue cities are biophilic cities, I have argued here. In the chapters that follow I tell some of the many stories of coastal cities designing with this marine nature in mind. There are now many inspiring stories of cities connecting residents to this wondrous nearby marine world, but also effectively preparing for the serious perils presented by climate change and sea level rise. It is an exciting and daunting time for advancing this new important model of global urbanism that is both blue and biophilic.

## NOTES

1. "How to Improve the Health of the Ocean," *The Economist*, May 27, 2017.
2. For a review of this evidence see Timothy Beatley, *Handbook of Biophilic City Planning and Design*, Island Press, 2017.
3. Rachel Carson, "Help Your Child to Wonder," 1956.
4. Rachel Carson, *The Sea Around Us*, Oxford University Press, 1951. Carson herself spent many hours exploring the Maine coast where she owned a small cabin and spent her summers.
5. More details about the Biophilic Cities Network can be found at www. biophiliccities.org. Cities joining the network must indicate the ways they are already biophilic and goals and steps they plan to undertake in the future; they must select and monitor over time a certain number of biophilic indicators; and they must adopt an official proclamation or resolution stating intent to join the network and to aspire to being a biophilic city.
6. See Richard Louv, *Last Child in the Forest*.
7. Interview with Jane Lubchenco, at the University of Virginia, March 20, 2017.
8. "Iceberg About the Size of Delaware Breaks off Antarctica," NBC News, found at: http://www.nbcnews.com/science/environment/iceberg-about-size-delaware-breaks-antarctica-n782096
9. Robert M. DeConto and David Pollard, "Contribution of Antarctica to Past and Future Sea Level Rise," *Nature*, 531, 591–597, March 31, 2016.
10. http://www.un.org/pga/71/2017/02/13/press-release-preparations-for-the-ocean-conference/

11. Interview with Murray Fisher, at the Harbor School, Governor's Island, New York City, May 8, 2017.
12. Wallace J. Nichols, *Blue Mind: The Surprising Science That Shows How Being Near, In, On, or Under Water Can Make You Happier, Healthier, More Connected, and Better at What You Do*, Back Bay Books, 2014.
13. Mathew White, Ian Alcock, Benedict Wheeler, and Michael DePledge, "Coastal Proximity, Health and Wellbeing: Results from a Longitudinal Panel Survey," *Health and Place*, 23, 97–103, September 2013.
14. Mathew White, Benedict Wheeler, Stephen Herbert, Ian Alcock, and Michael DePledge, "Coastal Proximity and Physical Activity: Is the Coast an Under-Appreciated Public Health Resource?" *Preventive Medicine*, 69C, pp. 135–140.
15. Russell Crawford, "The Impact of Ocean Therapy on Veterans with Posttraumatic Stress Disorder," undated, found at: http://amazingsurfadventures.org/asa/wp-content/themes/asa/pdf/Crawford-Report-Summary.pdf
16. The film can be found here: http://amazingsurfadventures.org/programs/operation-surf/
17. Murray Fisher, on-camera interview and site visit, The Harbor School, New York City, May 8, 2017.
18. For example, see Wilson, Biophilia, Harvard University Press, 1984.
19. "17 Goals to Transform Our World," found at: http://www.un.org/sustainabledevelopment/#

# CHAPTER 2

# Planning for the Balance of Danger and Delight

**Abstract** As a species, we are drawn to water and to marine edges and we gain many benefits from the blue, including wonder, enjoyment, stress reduction and greater meaning in life. Yet there are corresponding dangers associated with proximity to water, and climate change and sea level rise represent serious physical (and social and economic) design and planning challenges. The vision and practice of blue biophilic cities understands the need to balance the danger and delight: not only to provide new connections to water, but also creatively seek to adapt in ways that will make a city more resilient in the face of these physical forces.

We are drawn to coastlines and to the wonder and solace of beaches and marine environments. As argued earlier, proximity to water connects with our evolutionary past, it delivered considerable evolutionary benefits and it is no surprise that we find healing power in activities such as surfing, snorkeling and beach combing, or just sitting and watching and listening. Yet the paradox today is that especially with accelerating sea level rise these areas are increasingly hazardous as well. Increased flooding and the impact of coastal storms are ever-present dangers that need planning and adaptive design.

Many of the best emerging blue cities understand the need not only to enhance and expand access to the medicine and wonder of the sea but also

T. Beatley, *Blue Biophilic Cities*, Cities and the Global Politics of the Environment, https://doi.org/10.1007/978-3-319-67955-6_2

to work to minimize long term exposure of people and property to harm in extreme events such as hurricanes and long term shoreline erosion. Cities such as New York have emerged as wonderful examples of places that are rediscovering their waterfronts and waterscapes. This has happened through spectacular new waterfront parks, such as Brooklyn Bridge Park, and by opening up new vistas of the water. In New York, thanks to the concerted advocacy of groups like the Waterfront Alliance, there has been a major expansion to the ferry service, providing routes to underserved neighborhoods and for many residents significantly reducing travel times to the city.

We watched and experienced directly the advantages of the new ferry service as we traveled to meet Roland Lewis, founder and director of the Waterfront Alliance at Brooklyn Bridge Park. The trip from midtown Manhattan was close to an hour. The return was a quick five-minute ferry crossing on the city's new ferry service. Lewis is understandably proud of his role in creating the new NYC Ferry, which has doubled the number of neighborhoods in the city now served by ferry. This is good investment, much less expensive than rail or bus: and "so people in transit-starved neighborhoods such as the Rockaways or Soundview up in the Bronx can get to work or pleasure for a reasonable fare at a reasonable speed."[1] (Fig. 2.1).

Partly the strategies of New York are intended to ensure that enjoyment of the water, and the benefits of water, are enjoyed equally by all residents of the city. Mayor Bill de Blasio was one of the keynote speakers at the city's premier water event—the Waterfront Conference held each year on a boat (the Hornblower Infinity), organized by the Waterfront Alliance. De Blasio noted that day the high levels of ridership already seen on the new Rockaway to Wall Street route and declared that a citywide ferry service "can redefine how people get around. It's got limitless potential."[2]

There is still much to be done, de Blasio told the 500 or so attendees: "This city is surging forward. We're going to make sure the water is for everyone."[3]

An important part of the story of New York City—which has rediscovered its city motto of *City of Water*—is the power of organizing and of bringing together disparate groups and organizations around a powerful vision for the future of the city. The Waterfront Alliance shows what is possible to do in a decade. From its small beginnings as a project started under the umbrella of the Municipal Arts Society, it is now a large and robust coalition of some 950 organizations. About two years ago it further

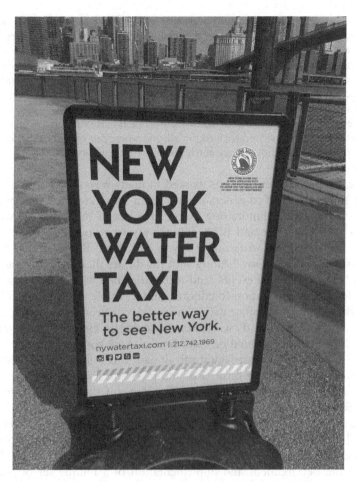

**Fig. 2.1** New York City is expanding its water transportation network in many ways, including through water taxis and by doubling its ferry service. Image credit: Tim Beatley

strengthened its brand, shortening its name from the Metropolitan Waterfront Alliance, to just Waterfront Alliance, and designing a new, visually striking logo.

The vision of this coalition is a powerful and compelling one, and one that could be embraced by other marine cities around the country and the world.

We believe the harbor and waterways of New York and New Jersey should reflect the vitality and diversity of the great metropolis that surround them. We envision a harbor and waterways alive with commerce and recreation: where sailboats, kayaks, and pleasure craft share the waterways with commuter ferries, barges, and container ships; where parks and neighborhoods are connected by affordable waterborne transit; where exciting waterfront destinations reflect the vitality and diversity of a great metropolis; where the waterfront is no longer walled off by highways and rails or by private luxury residences, but is a shared resource for all; and where our coastal city—a city of islands—intelligently and resolutely prepares for the reality of sea level rise.[4]

The Waterfront Alliance has worked in many ways to raise awareness about the harbor and to generate enthusiasm for re-connecting to water. There is the annual fundraising event, "Heroes of the Harbor," which occurs every October and is an opportunity to recognize some of the incredible leaders and work under way in New York. Recipients are given a "personalized ring buoy." Along with the dinner there is the "Parade of Boats." These types of events tend to demonstrate what Lewis argues is the role of his organization to educate, nudge, cajole. He points out that few mayors come into office with a harbor or waterfront plan. Groups like the Waterfront Alliance, Lewis believes, provide an essential "civic voice" about how the water should be used. As New York continues to transition away from its gritty, industrial waterfront past, this voice is greatly needed: "we might replace those businesses and piers and chain link fences with another series of barriers, like luxury buildings if there wasn't a strong voice for access, water quality, and jobs too."[5]

There are e-newsletters (the *WaterWire*), an annual harbor festival every July called City of Water Day, a harbor camp for kids that runs during the summer and, of course, a lot of policy and advocacy work. There is also the development and implementation of a unique set of coastal design guidelines, affectionately known as WEDG (Water Edge Design Guidelines).

In May 2017 the Waterfront Alliance issued its first "Harbor Scorecard." There is some good news here—many urban locations have access to water, and more than three-quarters of the water samples taken passed Environmental Protection Agency (EPA) safe swimming standards. On the other hand, the city still discharges more than 17 billion gallons of raw sewage each year as a result of its combined sewer overflow problem. And the number of residents likely to experience flooding is considerable, with a high percentage of these residents being "economically and socially vulnerable."[6] New York's scorecard in many respects encapsulates the current conditions of many blue cities—improving water access and connections, improving water quality, but with troubling pollution and vulnerability remaining (or even increasing, in the case of sea level rise).

Living and working in New York today is quite different than just a few decades ago, as the city rediscovers its waterfront. Central Park has been described as an effort to create nature as far away as possible from the edges of Manhattan, recognizing the gritty, grimy environment that was the working waterfront and not necessarily a place to relax or seek out for recreation (Fig. 2.2).

Many other cities are similarly seeking to re-establish connections to the water, to gain some of these blue mind benefits. Seattle, Washington, is another city where we have been filming and is another impressive blue urbanist story. Here the city is taking down the viaduct—the city's elevated highway, long serving to physically separate the city from its bay and sound. In its stead, the city is developing a 26-block park along Elliott Bay, a plan developed by James Corner of Field Operations and advocated by the nonprofit Friends of Waterfront Seattle.

The plan involves rebuilding Pier 62 and replacing the Elliott Bay seawall, and much of this has already happened. It is a "once-in-a-century" chance, says Friends of the Seattle Waterfront, "to create a park that will physically and psychologically reconnect us to our urban shoreline and Elliott Bay."[7]

Heidi Hughes, executive director of Friends of the Seattle Waterfront, spoke with me about this audacious plan as we walked along the beginnings of this new waterfront promenade. It will add some 20 new acres of waterfront public space and represents a return of residents to the waterfront, long a place more for tourists and visitors. Much of the work will involve improving intersections and making better pedestrian connections with the gridded streets that will feed into the waterfront. Hughes describes the vision of the waterfront as a kind of amphitheater, facing the city towards Elliott Bay and Puget Sound. "It's a real pivot from the city putting its back to the waterfront. Seattle will now reclaim its identity as a waterfront city."[8]

And it is a chance to do things differently and to experiment. In rebuilding the Elliott Bay seawall it has been re-imagined as a surface that will support a diversity of marine life, with special crevices and indentations to serve as new homes for algae and small marine organisms and the seawall

Fig. 2.2 Brooklyn Bridge Park, shown here, is one of New York City's new waterfront parks, and illustrates the city's new emphasis on physical and visual connections to water. Image credit: Tim Beatley

itself will serve as a salmon movement corridor. This is a new way to think of a seawall—an opportunity to enhance marine life at the same time as protecting against predicted sea level rise. "So you can imagine this whole park becoming a laboratory for families, teachers, kids, to reconnect with the shore, the ocean, and we can really expand what the Seattle Aquarium is doing throughout the waterfront." An expansion of the aquarium is also part of these ambitious plans.

Hughes makes a major point about the accessibility of this new park— like many other parks in the city, it will be easily reachable by all residents. "This will be the most accessible park for the whole city." And that makes sense—it is the water, and the extensive watery edges, that will bind a city. It will be easily accessible by transit and there will be new dedicated bicycle tracks connected to a regional network of bikeways.

Access to urban amenities and economic opportunities in cities has not always been fair. Equity issues of course loom large in some of these places. Baltimore is another example of a harbor city seeking renewal and restoration, though not without the challenges of crime, ethnic conflict and economic inequalities. How do all residents of the city enjoy the benefits of water and the jobs and income derived from it? We spent a September morning with kids from a nearby disadvantaged neighborhood who were learning how to kayak under a special program run by the city's Parks and Recreation department. The goal was to get kids on the water and to allow them to experience first hand this different place. First they sat in the kayaks on land, then eventually, and with guidance and encouragement from Parks and Recreation staff, they carefully navigated around in the water. As we learned that day few of these kids had ever been on the water, even though they lived in neighborhoods not far away.

Baltimore's inner harbor has the distinction of being an important proving ground for demonstrating cultural return to the water. Visionary developer James Rouse built one of the first successful commercial harbor projects—Harborplace—here some four decades ago, setting an example for other cities and demonstrating that indeed there was a desire for people to spend time, to visit, to play along urban waterfronts. A remarkable plan for the inner waterfront, prepared by Wallace, Roberts and Todd in the 1960s, made much of this possible and helped set an example for other cities to plan for a re-discovery of their waterfronts. These early Baltimore visions are now finding new expressions and new manifestations.

One of the main advocacy groups in Baltimore is the Healthy Harbors Initiative run by the Waterfronts Partnership. The partnership is itself a coalition of different local organizations with a stake in the harbor. It runs essentially as a business improvement district (BID), but with an unusual mission: serving as the harbor's "chief advocate, promoter and steward." This is on top of the other more typical services provided by a BID (such as litter control, visitor assistance and landscape management).

One key member of the partnership is Baltimore National Aquarium. As a pre-eminent urban aquarium, it is re-thinking its mission in the urban age. With the help of Chicago architect Jeanne Gang the aquarium has launched a plan for its re-design and expansion. The new plans will better connect it to surrounding water and will include an urban wetland as a key feature and focal point.

Most of the focus of the Healthy Harbor Initiative has been on reaching the goal of a swimmable and fishable harbor. The intent is to reach this goal by 2020, and they are a long way off. Yet there is creative work being carried out and good progress being made. Among the creative steps taken so far are the installation of floating wetlands in a number of places, and support for the growing of oysters in the harbor, a crop that helps to clean the water.

One of the projects that has gained the most attention is one with a humorous slant—the design of a one-of-a-kind inner harbor water wheel. Affectionately dubbed "Mr Trash Wheel" it is a structure that sits on the water in the harbor with a rotating water wheel that is turned partly by solar power, but mostly by the river current. It scoops up a remarkable amount of trash and debris: some 182 tons so far (including bottles, plastic bags and more than 4 million cigarette butts). You can follow Mr Trash Wheel on Twitter (with the handle @MrTrashWheel he has more than 12,000 followers!). In the latest development, the solid waste collected and prevented from entering the harbor is being used as fuel to generate power in a local waste-to-energy plant (Fig. 2.3).

Growing oysters in a city harbor is a clever way to engage the population. The young oysters (spat) are found in cages in the water at prominent locations around the harbor, tethered by ropes. Some cages are sponsored by schools, others by local companies or corporations, and periodically volunteers show up to pull up the cages and clean them.

We watched Adam Lindquist, who heads the Healthy Harbor Initiative, pull up several cages and immediately a small crowd of tourists formed. They appeared fascinated by what appeared mysteriously from the shallows and proceeded to pepper Adam with questions. It was a visceral demonstration of just how interested people can be and how potent a tool

**Fig. 2.3** Mr Trash Wheel, shown here, is located in the inner harbor of Baltimore, Maryland, where it utilizes river current and solar energy to collect waste and debris. Image credit: Adam Lindquist, Healthy Harbor Initiative

oysters could be in educating and inspiring about a healthier blue future. I would learn later about an even more ambitious effort—the Billion Oyster Project—run by the New York Harbor School (and discussed in some detail in Chap. 5), showing how urban residents and harbor visitors could find excitement and fascination in the watery nature around them.

Planting oysters helps to clean our marine waters and it can also serve as a form of adaptive response and mitigation tool in the face of sea level rise and coastal storms. Landscape architect Kate Orff has been pioneering the idea of "oyster-tecture," utilizing oysters and oysterbeds in cities like New York to enhance habitat, serve as protection barriers to storms and flooding, and to engage the public. Her ideas are now being applied in a large way along the southern shore of Staten Island.

How to adapt to sea level rise and climate change in Blue Biophilic Cities remains a major challenge and an open question. But we are finding some emerging models for guidance and inspiration. Increasingly, cities recognize the value in maintaining and enhancing natural systems to respond to sea level rise. There will be the chance to expand and enhance natural habitats at the same time that more dynamic and resilient shorelines are designed and installed.

Cities such as New York are investing in innovative projects including Orff's Living Breakwaters project and the so-called Dryline, another project funded following hurricane Sandy. As these new approaches demonstrate, every adaptive response and investment should seek to advance multiple goals—not just achieve flood mitigation, as a seawall or revetment might, but also provide recreation, habitat restoration and new visual and physical connections to water and the marine realm. And perhaps other goals as well—the generation of renewable energy, for example, or eventually the sustainable and local production of seafood.

Some Blue Biophilic Cities will have a harder time with climate adaptation than others, and here Miami Beach stands out. It is a city with a special position on the very front line of climate change and sea level rise. We recently spent time in this barrier-island city speaking with officials about what the future holds there and the efforts being made to effectively adapt to long term sea level rise.

Many coastal mayors and leaders are in denial about sea level rise. This has not been the case in Miami Beach. The current mayor, Philip Levine, even campaigned on the issue, famously appearing in a commercial (with his boxer dog Earl) paddling through flooded streets in a kayak. Levine declares proudly that he was not swept into office but "floated into office."

Cities like Miami Beach will definitely have their work cut out for them. The severity and swiftness of sea level rise that coastal cities will experience is still unclear and hotly debated, but as ice sheets melt more quickly than expected, the prognosis is not optimistic. One scientist who has been vigorously sounding the alarm is Professor Harold (or Hal) Wanless, Professor and Chair of the Geology Department at the University of Miami. Wanless points to the NOAA 2012 National Climate Assessment predicting global sea level rise of between at least 4.1 and 6.6 feet by 2100.

He argues that we are clearly in one of those "very rapid pulses" of accelerated sea level rise seen following the last ice age about 18,000 years ago. South Florida has already seen a foot sea level rise since 1930, so this is anything but hypothetical. Wanless describes the way that oceans are absorbing global heat, leading to rising sea temperatures which in turn perilously penetrate and melt ice sheets and global fjords, leading to a damaging positive feedback loop.

"We probably should be anticipating at least 7–30 feet of global sea level rise by the end of the century regardless of what we do. Even if we stopped burning fossil fuels tomorrow, the greenhouse gases in the atmosphere will keep warming the atmosphere for at least another 30 years. More than 90 % of this global warming heat is ending up in oceans."[9]

What to do? Wanless says, among other things: "Terminate long-term, infrastructure-intensive development of barrier islands and low-lying coastlines…and divert public money from hard and soft shore-protection measures into funds for relocation assistance, clearing low-lying polluted lands, and removing storm-damaged development and infrastructure."[10]

And we need to do a lot more getting ready for these changes, a lot more planning. Wanless argues that we need to establish thresholds for saying no: for terminating financial commitments to provide services and infrastructure, and we need to prepare in advance of storms and hurricanes to say no to re-building in many dangerous places following future storms.[11]

What can a city like Miami Beach in fact do? What are the potential planning and design options open to them? The cities of South Florida actually have fewer options than other coastal cities to the north. Miami Beach sits on geology made up of porous limestone. That means building seawalls to protect this city will not work, at least not in the long run.

City engineer Bruce Mowry explained to me that much of the flooding seen here is what can be called "rainy day flooding," that is flooding that results not from a storm or high tide necessarily, but simply a result of backflow from the city's antiquated stormwater pipes (Figs. 2.4 and 2.5).[12]

**Figs. 2.4 and 2.5**   Miami Beach has been elevating roads in neighborhoods such as Sunset Beach, shown here. It creates an interesting physical disconnect between street level and storefront entrances, and sometimes, as shown on the right, a sheltered seating area for customers. Image credit: Tim Beatley

The city is implementing a $400 million program of investments in adaptive infrastructure, championed by its mayor, Philip Levine.[13] The first thing Mowry recommended doing (when the mayor told him he needed to eliminate rainy day flooding in a year) was to install new check values in some 70 locations in the city to prevent backflow. But this of

**Figs. 2.4 and 2.5**  Continued

course is not enough and the city has embarked on major effort to shift away from its inadequate and antiquated gravity based stormwater system to one involving pumps. The city will likely eventually have installed some 60 to 80 new pumping stations around the city, the precise number, Mowry tells me, is not yet known. He's moving the city towards pumps that are lower horsepower but higher capacity—able to pump some 80,000 or more gallons per minute. Around 20 of these pumps have been installed already, with another 12 in construction, so the city is about halfway through building the new network of pumping stations it needs.

Another key part of the city's short term strategy is to gradually elevate roads and sidewalks. The city has recently completed a major road elevation, of the order of 2.5 feet, in the neighborhood of Sunset Harbor and is beginning work on several other neighborhoods. Walking around the Sunset Harbor neighborhood the road elevation is obvious, as most store and restaurant entrances now require a pedestrian to step down. In several cases this has created not unpleasant new patio and seating spaces that are separated from the street above.

Seawalls are also similarly being raised. The city has floated bonds to cover the initial phase of these infrastructure investments and has raised household stormwater fees by a modest amount ($7 per household). We visited and filmed this neighborhood in January 2017. The elevation of the road meant that many of the shops and restaurants now had entrances that were below the street and had to be accessed from stairs leading down from the sidewalk and road above. In many cases this has lead to the creation of shielded areas of tables and seating, in the case of restaurants: not a bad result, actually, with a feeling of being more separate and protected from street traffic. If properly explained to residents and visitors these areas could also be a source of discussion about expected sea level rise and the city's efforts to plan for it.

The city is exploring other ideas as well. Mowry spoke of promoting the idea of "adaptable architectural features," especially in new commercial structures. One idea Mowry described was a kind of movable first floor, made possible by designing higher than normal ceilings. As sea level rises and as streets and sidewalks are elevated over time, the usable first floors of buildings would also be raised.

There is also a recognition that the city's beach re-nourishment efforts must continue. One of the nation's first coastal communities to re-nourish, or replenish it beaches, this has been a continuing and ongoing expense, but understood as necessary to continue to support its tourism industry. Visit Miami Beach's Ocean Drive on any day and you will see a vibrant and popular destination, with many of the qualities of great built environments—sidewalk life, high walkability and a creative mix of uses and activities.

While there is understandable stress and pessimism about the future of Miami and South Florida, there are many who are attracted to its closeness to the ocean. The beach (re-nourished as it is) is beautiful, and the nearby ocean is luminous and beautiful. The majesty and wonder of a remarkable marine biodiversity is not far away, indeed it is all around, as residents and visitors scan the seascape for signs of dolphins.

And there are remarkable efforts to connect residents and visitors to the abundant marine life that lives in the shadow of this high-rise coastal paradise. The Marjory Stoneman Douglas Biscayne Nature Center (described more fully in Chap. 5) is one such place that has been doing this for decades.

There is little illusion in Miami Beach that that these interim steps will protect the city in the long run. The mayor, city engineer and others have framed these investments as necessary adaptations that will carry the city 30 to 40 years into the future. Beyond that there is a sense of optimism about what new physical designs and technology might help the city. The city is not giving up and will not go away, that is the clear message of leadership there.

Mayor Levine is quoted in *Miami New Times* expressing this optimism:

> I know that human innovation is so incredible. If I told you 30 years ago that an iPhone could Facetime with a friend in Europe now in real time, you'd think I was out of my mind. The opportunity for entrepreneurs is unlimited. They'll come up with solutions we can't even think of today. Deep-water injection pumps below the aquifer? Who the hell knows.[14]

City engineer Mowry echoed these sentiments in his interview with me. "The goal is that we're going to be here forever...Retreat is not considered here as an option."

The city was formed from the swamp 100 years ago and it will find a way to sustain and reinvent itself even in the face of rising seas. "Why can't we for the next 100 years do the same thing but do it in a little different fashion? We're not going from a swamp to a city, we're going from a city to a different city and who says that our culture should be the same? And who says we shouldn't be more water-based, who says we may not have elevated transportation, who says we may not have buildings on piers that embrace the water, and that actually allow the water to come in, but we build structures over the water as a city?"

Serious retreat is framed as a giving up on the city. "We believe in our city." Engaging in retreat suggests we don't love our city and we are "willing to use it up and get out." Here we see both a failure to confront the realities of the magnitude of sea level rise and the sentiments of love and fidelity towards a place that is in many ways quite admirable.

We need to be open to many new ideas, says Mowry, and he places a lot of positive stock in design studios and students from Harvard Graduate School of Design and Florida Atlantic University who are imagining creative design solutions to this high-water future. He tells them, "don't look at boundaries."

Not far away in Coral Gables, home to the University of Miami, there is a similarly daunting sea level challenge. Here a quarter of the city's property value lies below 4 feet in elevation. Here we met and filmed the Republican party mayor of Coral Gables, Jim Cason, who, like Levine, is not in denial but seeks to understand and get ready for sea level rise. This is a city with a history of building and living in ways intimately connected to water, including an original network of Venice-like canals.

Coral Gables has a huge problem as well, with some $15 billion in property at risk. The steps this city has undertaken to be prepared and get ready include more detailed LIDAR maps showing the city's vulnerability; a careful assessment of critical facilities at risk; and even commissioning a paper that explores the legal aspects of hazard mitigation and retreat (as Cason tells us, to be clear what the city's duties and options might be "when your streets are underwater and it's too expensive [to rebuild them]").[15] Cason explains further, there are many things the city can do to reduce greenhouse gas emissions and the city is doing many of them through an ambitious sustainability plan, including shifting to electric vehicles, mandating LEED-certified construction, more energy-efficient municipal lighting, among other actions.

The city of Coral Gables is already taking steps to adapt to sea level. Cason tells of a recent bridge renovation where the structure was raised two additional feet. "We need to start making sure that everything we do in the future that has to do with construction or rehabing takes sea level rise into account," Cason told us.

Cason seems proudest of his efforts to raise awareness and start conversations about the sea level rise that is coming. "Physics is physics," he declared. "It's not a partisan issue, it's a fact. We see it everyday." This is refreshing candor from a mayor who sees addressing these challenges as an important leadership dimension.

"For us it's a reality. It's a leadership issue. It's an obligation. We've all got grandkids. We've got to tell them not only what's going to happen but also what they can do in their individual lives." A relatively small community, with around 50,000 permanent residents, Coral Gables shows some of the essential steps blue cities can and must start to take, beginning with information and candid community discussion about what to expect down the road.

These sea level rise challenges are being faced by many coastal cities around the world, of course, and international examples of big and small cities, provide additional hope and ideas. Rotterdam is one coastal city, in a nation that has a long history of keeping back the sea, that continues to offer inspiration. Its goal is to be "climate-proof," by 2025, and it has undertaken an unusual array of strategies to address the management of flooding and water. In an interview, Chief Resilience Officer Arnoud Molenaar emphasized the approach of "no-regrets" solutions: essentially steps that could be taken, projects designed and built that would be beneficial even if not protective or needed for climate adaptation. In the summer of 2016, I visited some of these. They include designing new parking structures to retain stormwater, and the installation of so-called "water plazas" that can serve to hold rainwater during times of storms and flooding, but would all other times serve as public spaces and community gathering spots. A visit to one of the first suggests these spaces are interesting and well designed. Here there are stepped edges, places for kids to play and a sunken basketball court (Figs. 2.6 and 2.7).

**Figs. 2.6 and 2.7**   Rotterdam has employed a variety of urban design strategies to retain and control water. Shown here is the Bethemplein—a so-called "water plaza" that serves to collect and retain rainwater, but on sunny days adds a public space to the neighborhood. Image credit: Tim Beatley

**Figs. 2.6 and 2.7** Continued

Many of the things envisioned by the city as a way to retain and control water are also biophilic. The city has encouraged and subsidized the installation of many green rooftops. Several of these I visited, including one large one.

The Dutch have famously pioneered the idea of "floating communities," where entire neighborhoods of new homes are able to rise and fall with the water levels. Rotterdam has envisioned an entire floating neighborhood in the inner port, and has installed model structures that show what is possible when designing with special lighter building materials.[16]

And there is also the now-famous DakPark (or roof park) in Rotterdam, essentially a large park, on top of shops and businesses. It is in a sense a very large green roof and a quite beautiful, if slightly unusual, municipal park. The city has encouraged and subsidized the installation of green roofs with considerable success: there are now more than 220,000 square meters installed. The DakPark is a dramatic example but there are many other places in the city that sport such resilient, biophilic design features. These are steps similar to those taken by the Dryline in New York that demonstrate clearly that it is possible to live in a green and biophilic city, to advance a vision of biophilic urbanism that will also address the long term dangers.

## Some Concluding Thoughts

Coastal cities, blue cities, face at once daunting challenges and exciting opportunities. Climate change and sea level rise will represent ominous threats to be sure. But many cities, from New York to Seattle, recognize the possibility of taking design and planning steps that will adapt and respond to future sea rise and coastal flooding, but will also see that humans are drawn to water. New water-oriented parks and shoreline and harborfront public spaces, such as Brooklyn Bridge Park, reflect the increasing desire to be near, on or in the water. It is the premise of this chapter (and book) that cities can respond to both impulses, both challenges.

The ability to see and be near water, to enjoy marine environments, responds to a deep biophilic need and we know that we will be happier, healthier and lead more wondrous, meaningful lives through daily contact with the marine nature around us. In short, Blue Biophilic Cities revel in the delight and magic of the marine world, but also must plan for and adapt to the increasing dangers associated with proximity to water.

## Notes

1. Personal interview with Roland Lewis, president and CEO, Waterfront Alliance, May 9, 2017.
2. "Nuggets of Wisdom at the Waterfront Conference," Waterfront Alliance e-newsletter, May 12, 2017.
3. Ibid.
4. "Our Vision," found at: http://waterfrontalliance.org/who-we-are/about-us/
5. Personal interview with Roland Lewis, Brooklyn Bridge Park, May 8, 2017.
6. Waterfront Alliance, "Harbor Scorecard," found at: http://waterfrontalliance.org/what-we-do/harbor-scorecard/
7. Friends of Seattle Waterfront, Annual Report 2016.
8. Personal interview, Heidi Hughes, Executive Director, Friends of Seattle Waterfront, January, 2017.
9. Harold Wanless, 2015, p. 2.
10. Wanless, 2015, p. 4.
11. Wanless says ominously, "Without planning, there will come a point in society and civilization as we know it will collapse into chaos." Ibid.
12. Interview with Bruce Mowry, City Engineer, Miami Beach, Florida, January 25, 2017.

13. Jessica Weiss, "Miami Beach's $400 Million Sea Level Rise Plan is Unprecedented, But Not Everyone is Sold," *Miami New Times*, April 19, 2016.
14. Weiss, Jessica, "Miami Beach's $400 Million Sea-Level-Rise Plan is Unprecedented, But Not Everyone is Sold," *Miami New Times*, April 19, 2016.
15. Personal interview with Mayor James Cason, in Coral Gables City Hall, January, 2017.
16. For more about this see Beatley, *Blue Urbanism*, Island Press, 2014.

# CHAPTER 3

# An Unsustainable Bounty from the Blue: Cities to the Rescue?

**Abstract** What (and how) we grow, harvest or extract from the ocean has significant implications for long term sustainability of this immense ecosystem. Our current industrial approach to seafood harvesting is clearly not sustainable and cities can and must take the lead in developing new approaches. Urbanites must begin to shift their consumer and political power behind more sustainable ideas and practices. Some of these new approaches are explored here, including support for smaller scale, locally based fishing (and new mechanisms such as Community Supported Fisheries [CSFs]), and a shift towards more sustainable and humane forms of shellfish aquaculture and ocean vegetable farming.

The ocean has been an immensely important storehouse (or stock of assets) for the human species. We pull many things from the ocean (or from under the ocean), oil and gas, sand, minerals and of course we harvest seafood. None of these is especially sustainable and, as recent major oil spills suggest, drilling for oil is not a wise form of stewardship for planetary oceans. We have depended on the global bounty of seafood for millennia and, in theory at least, it might be possible through careful limits and management to harvest seafood in a sustainable way. That is not the reality in most parts of the world and the prognosis for global fisheries is not very encouraging.

© The Author(s) 2018
T. Beatley, *Blue Biophilic Cities*, Cities and the Global Politics of the Environment, https://doi.org/10.1007/978-3-319-67955-6_3

Few marine biologists have been as creative in understanding the human impact as Daniel Pauly, who heads the Sea Around Us Center at the University of British Columbia in Vancouver. I interviewed him in January 2017 about his groundbreaking work attempting to more accurately understand the extent to which we extract fish from the ocean. His numbers are alarming. Engaging in something he and his colleagues call "catch reconstruction," it is a painstaking effort to more accurately and fully account for all the fish extracted from the sea.

Pauly's conclusions suggest that we have extracted more than we previously thought—the global seafood catch likely peaked, he believes, in around 1996 at 130 million metric tons, significantly higher than the (official) estimates of global catch put forth by the Food and Agriculture Organization (FAO) of the United Nations (UN). And even more alarming the decline in total global catch since that time has been more rapid than previously thought.

Why the sharp decline in fisheries globally? Pauly is especially critical of what he refers to as the "fishing industrial complex." Analogous in many ways to the rise of agribusiness and a corporate land-based farming system, we are increasingly subsidizing a fishing industry that employs increasingly larger ships and equipment to go after increasingly farther-out, deeper sea species. Pauly estimates the subsidy of the US fishing industry at $30 billion annually.

He likens the whole system to a "giant ponzi scheme." There are many more fishing vessels, he says, than the marine resource base can sustain or support, and they simply shift and move to other, more marginal and more remote fisheries when others become commercially exhausted.

Part of the challenge of responding to diminished fisheries is a phenomenon that Pauly coined, "shifting baselines." It explains why we seem not be as alarmed as we should about the catches we're seeing in certain fisheries. We compare the decline we see to near term metrics and fail to compare with longer term, historic conditions that would show more clearly how much we have lost. The longer timeframe might be (should be) hundreds of years, but we are likely to compare changes with those seen only over the course of one's lifetime.

Stronger regulatory systems and more stringent management of fisheries would accomplish much. Jane Lubchenco, former NOAA administrator, speaks optimistically of the potential to strengthen management and the prospect of re-building and restoring depleted fisheries. We need to "fish smarter, not harder." She offers the USA as a case in point, seeing a number of fisheries experience "remarkable turnarounds." She offers the example of the west coast groundfish fishery, a fishery that includes some 90 different species, including cod, rockfish and whiting). A "catch shares" system, implemented in 2011, has led to its remarkable recovery. Changing the economic signals and giving fisherman a direct personal stake in responsible management can lead the way. Lubchenco is encouraged, moreover, by new, innovative approaches (Fig. 3.1).[1]

How quickly could we restore fisheries? Quite quickly Lubchenco says, in as little as 10 years. This dose of optimistic thinking is necessary, and the development of more stringent and innovative fish management regimes is essential. Cities and city political and environmental leadership can push and agitate for these new rights-based systems. But the global trends and pressures will make it hard to bring about this scenario of restored, replenished and sustainable global fisheries. The expansion of industrial fishing, the growing size and range of boats and the increasing destruction wrought by technology, coupled with a rise in global demand for fish, make the rosier scenarios less realistic. And even the best management regimes will face new challenges, such as the shifting movement of fish stocks in response to warming waters.

Pauly and Lubchenco agree on the importance of expanding marine protected areas and that must be one part of the answer as well. Lubchenco notes the positive trends here—on a global scale marine protected areas amount to only 2 % of the surface of the planet, yet we have seen this increase dramatically in a short period of time (from less than 0.2 % a decade ago). Thanks to the leadership at federal level, and the personal support of former President Barack Obama, the extent of US waters in marine protected areas has grown considerably, now at about 23 %. What should this figure be globally? Coastal nations around the world have committed to protecting 10 % of their respective ocean areas by 2020— this is a good start, certainly.[2] Lubchenco and other marine conservations say it will take at least 30 %, but perhaps in line with E.O. Wilson's half-earth concept the percentage should be even higher.

The other good news is that research clearly shows that fish stocks recover in no-fish areas, and marine protected areas can play a significant role in helping to re-stock fisheries. There is also evidence that marine ecosystems, including coral reefs, tend to be more resilient (less likely to experience bleaching) when embedded in larger marine reserves.[3] Urbanites and cities must support and lobby for the designation of larger protected areas.

**Fig. 3.1** The Pikes Place Fish Market is an iconic institution in Seattle, Washington, and a popular tourist destination. The market is now only selling fish that are sustainably harvested. Image credit: Tim Beatley

Much of the work (and legal authority) of establishing new marine protected areas falls of course to national governments. Nevertheless, cities and states can do much to advocate for them and connect them to efforts at establishing more local marine parks and integrated land–sea parks or "bluescapes." California has a remarkable story, one to be very proud of, establishing an extensive network of marine protected areas through the Marine Life Protection Act (MLPA), and adjacent cities and communities have now adopted them as their own, taking active steps to monitor and enforce the protected areas. "It is possible to use the ocean without using it up," Lubchenco says, and the California story makes this clear. Many cities may have the chance to establish marine protected areas very close to where residents live. Indeed, the new waterfront or ocean-front park must be re-imagined as an opportunity to integrate land and sea and to protect important nearshore marine environments, many of which have been negatively impacted by terrestrial activities such as vegetation clearance and non-point-source pollutants (Fig. 3.2).

Any trip to Seattle or to San Francisco might not feel complete without eating seafood, and there are many ways that the choices of seafood we consume can make a significant difference.

There are now efforts to certify sustainably managed fisheries, for instance, through the highly successful Marine Stewardship Council, and there are many efforts by restaurants and others to source and feature more sustainable seafood choices. The San Francisco Bay Area has become an epicenter for these efforts, especially through the work of the Monterrey Bay Aquarium's Seafood Watch program. The aquarium used to publish small cards (I kept one for many years in my wallet) providing guidance on which fish and seafood to avoid buying, either because of the condition of the fishery or the methods used to catch the fish. Now this information is mostly delivered through a free Seafood Watch app.[4]

We visited and filmed a San Francisco restaurant that has been highly involved in actively sourcing and selling more sustainable seafood. On a day in March we sat down with Bob Partrite CEO and manager of the Fog Harbor Fish House Restaurant, an established seafood restaurant located on Pier 39 in the heart of the San Francisco waterfront. The messaging to patrons here begins with the menus which state clearly the importance of sustainable seafood. Partrite clearly has a passion for sustainable seafood and took personal pride in telling us what they serve and the steps he and his staff take to ensure responsible sourcing of what they buy. They try to follow religiously the standards and guidance of Seafood Watch. At the

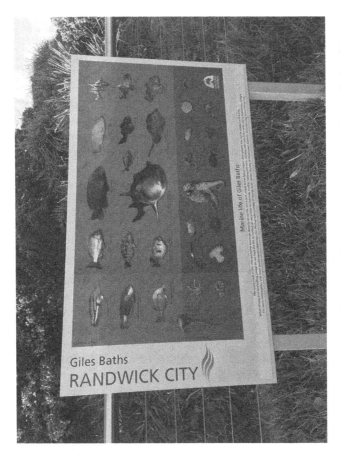

**Fig. 3.2** Along the edges of Sydney's South Beaches there are many opportunities to find and interact with marine life. Shown here is one of the many signs erected by Randwick Council to help residents look for and identify some of the more common marine species they are likely to see. Image credit: Tim Beatley

end of our interview, Partrite took me back to the kitchen to show me the "white board," a pretty impressive and complex listing of all the fish and seafood served, with specific information about its sourcing. I could immediately see the value of this and how working there it would be easier to answer the questions concerned customers might have. I could also get a strong sense of how empowered (proud even) I might be working at this restaurant (Fig. 3.3).

On that day and during our conversation with Partrite we ordered a few items off the menu, including the Dungeness crab, an important locally sourced seafood. Partrite described in detail how it was fished and the condition the fishery was/is in, and the fact that little or no bycatch resulted from the harvesting of this seafood. I also ordered a grilled salmon salad and the story here was a bit more complicated. There is wild caught salmon available some of the year, Partrite told us, but not at the moment. The salmon that arrived on my salad was from a salmon farm in Denmark, but one where the fish are raised in a closed-circle, recirculating system, with fewer of the negative impacts associated with open-ocean salmon farms.

The interview that day was both optimistic and reassuring about the potential power of consumers, and consumers educated and led to better choices through passionate restaurateurs like Bob Partrite. However, it also reinforced just how difficult it is to get sustainable seafood right.

The Fog Harbor Fish House Restaurant is part of a larger consortium of restaurants, managed by the Aquarium of the Bay. Carrie Chen, director of education and conservation at the aquarium spoke about the initiative and how it includes some 30 restaurants.

One way we are responding is through expanding aquaculture; but all aquaculture is not the same and it is alarming that there is not a better general awareness (for instance by restaurant owners and servers) about how ecologically damaging much aquaculture can be. One recent experience ordering seafood at a restaurant in Florida was telling. I asked about the "Scottish salmon" on the menu—what did the waiter know about its provenance, and specifically was it wild caught? I suspected it wasn't, but was surprised at what the waiter said on returning to the table: "It's kind of like wild because they're raised in pens in the water."

A month later I met with the pre-eminent fisheries ecologist, Professor Daniel Pauly, from the University of British Columbia. Pauly's work shows that we suffer from what he calls "shifting baselines"—that is, we judge and perceive the status of things like fish abundance through a very limited time lens.

**Fig. 3.3**   The Fog Harbor Fish House Restaurant in San Francisco has committed to serving only sustainably harvested seafood. Here the general manager show the author the so-called "white board" in the kitchen, which summarizes and keeps track of all the seafood purchased and served in the restaurant. Image credit: Tim Beatley

This unfortunately is the kind of aquaculture that Pauly calls "carnivore farming": highly damaging and like fattening cattle it is not very efficient in the use of limited fish catches. "It takes...pounds of [wild fish feed]... to produce one pound of salmon," says Pauly. Pauly supports stronger catch quotas and more marine protected areas where fishing is limited.

It is clear that our collective knowledge about seafood and fisheries is scant and indeed our ocean literacy overall is disturbingly inadequate.

Pauly argues that we must reduce the overall global harvest of fish and that, counterintuitively perhaps, will lead to increased fish production. We have to fish less to catch more, he says. We need to allow stocks to rebound, through stricter quotas and more extensive no-fish zones.

Still he says, "humanity cannot depend on fish," to feed itself. We should begin to consume fish further down the food chain—instead of the inefficient practice of fishing species like Peruvian anchovies or mackerel to feed other fish, we should return to directly consuming them (perhaps re-kindling an appetite for such smaller, but flavorful fish).

Pauly spoke of the possibility of a shift in how we see eating fish, the possibility at least in the wealthier North of moving away from seeing fish as a staple and rather more of what Pauly calls a "ritual food," something we eat only occasionally and perhaps only at special events or occasions. Would we reduce the fishery-exhausting, damaging consumptive patterns, while also infusing greater meaning to the fish we do eat? It is hard to know, but given declining global catches and the need to protect local fisheries for local people, movement away from a staple of fish seems to be something in our future.

Pauly offers a somewhat hopeful vision about a more sustainable rela-tionship with the fisheries around where we live. "I would like to dream of an economy where we are all locavores," Pauly says. What he has in mind may be something analogous to what we have seen around the slow food movement. Perhaps "slow fish" could emerge as an equally attractive idea, that we seek to learn about the provenance of the fish we eat, we support local fishermen and suppliers and we relish and savor the taste and enjoyment that comes from eating more locally sourced fish with a deeper story.

Part of the answer must be new mechanisms for even more directly connecting local urban consumers with local and regional fishermen. There is much benefit to be had on both sides. New mechanisms like CSFs hold promise. Modeled after the land-based Community Supported Agriculture (CSA) in which residents buy shares that entitle them to a weekly allocation of fish, this mechanism connects the fishermen directly with the consumer. That has the potential to promote learning about the fishery and to foster a sense of emotional connection with fishermen, and also (like a CSA) helps fishermen financially—more of the profit from fishing is kept by the fisherman, and the urban consumer helps to finance and underwrite some of the risk involved in a fishing operation.

One CSF that operates along the Californian coast in the small fishing village of Moss Landing, is the Real Good Fish CSF, co-founded and run by Alan Lovewell. On a visit there we also had the chance to meet and interview a local fisherman, Calder Deyerly, who sends much of his catch through the Real Good Fish CSF. Later in the day watched as he unloaded what he had caught earlier. It was quite a contrast to the image of industrial fishing with mammoth boats and legends of workers. It was just Deyerly working alone. Later he received a little help from dock crew, lifting the several relatively small crates of fish he brought back (Figs. 3.4 and 3.5).

**Figs. 3.4 and 3.5**   Calder Deyerley, shown here, is proud of his family's fishing heritage. He fishes along the Californian coast near Monterey using sustainable techniques, and sells his fish to local restaurants and to the local CSF. Image credit: Tim Beatley

**Figs. 3.4 and 3.5**  Continued

Here was a scale where you could see and comprehend everything and you could personally get to know your fisherman as we did that day. Deyerly had nothing to hide, was proud of what he was doing and spoke passionately of the fishing heritage he inherited from his fisherman father and that he hoped to pass down to his own 5-year-old son. This was the face of a fishing industry that we could support and be proud of, I thought to myself.

On this day he was fishing sablefish (also known as black cod) on a long line. He returned in the afternoon having reached his quota.

There is little doubt that support for local fishermen like Deyerly would lead to more sustainable fisheries. There is a connection and accountability here, and even the chance for urbanites to learn about the ocean from these interactions. The emerging CSF mechanism is an important one, but still remains small in size. There are perhaps a couple of dozen CSFs scattered around the USA.

This is perhaps not surprising given that our national policies and systems of subsidies tend to work against these kinds of small scale fishermen. Without subsidies the deep sea trawling that we see today would not be economical, Pauly tells me.

Indeed the larger trends suggest ever greater support for the industrial forms of fishing that are so ecologically destructive and human–ocean disconnecting. A recent analysis of the financial subsidies for fishing finds (not surprisingly) that the lion's share goes to larger, industrial fishing operations. Writing in *Marine Policy*, Schuhbauer and co-workers conclude that of an estimated $35 billion in national financial subsidies given to support fishing (in 2009), only about 16 % went to the "small-scale fishing sector."[5] The parallels to land-based agriculture are evident. As a matter of fairness this ought to concern us, but more important is the resource-exhausting, unsustainable industrial scale fishing supported by these subsidies.

One interesting development is the growing awareness of the emotional complexity of fish, adding an important and relatively new dimension to the debate about seafood harvesting. Could there be another moral reason to shift entirely away from fish, built less on scarcity and unsustainability and more on sentience and the need to avoid pain and suffering?

New research is adding to our appreciation of the emotional and psychological complexity of fish and a sense of concern about how we are treating these creatures literally swept up in the industrial fishing system. In many ways the emerging concerns parallel the concerns we feel about the plight of animals in land-based factory farms, and while considerable progress has been made in improving animal living conditions and introducing a measure of humaneness, no similar efforts can be seen (yet) when it comes to fish.

Much of our early awareness of the plight of animals and the need to take their pain and suffering into account can be traced to Australian philosopher Peter Singer, and to his groundbreaking book *Animal Liberation*.[6] I spoke recently to Singer about progress made since the book was published in 1975. He is largely optimistic and points to progress in improving conditions for farm animals, especially in the European Union, but points to the marine realm as an area where little has been done so far.

There is now good research showing the psychological and emotional complexity of fish, something not heretofore given much attention. This research shows, Singer believes, that "fish are individuals...some are quite intelligent, good at problem-solving...they have emotions and feelings, and relate to others as individuals."[7]

By Singer's estimate the annual global fish kill exceeds a trillion: "It is an enormous number of sentient beings being killed...And of course many of these deaths are very painful and drawn out. So I think it is time that people become aware of the slaughter, of the suffering involved," as well as the unsustainable and damaging nature of modern fishing methods. This is a lot to consider and an issue not widely thought of when ordering a salmon or grouper for dinner.

There are other reasons why we have been slow to care about the pain and suffering of fish. "They can't vocalize, or we can't hear them," in the same way that we might hear a pig or a bellowing cow. "Fish are silent to our ears," Singer says.

I must admit to my own visceral emotional reaction to the landing of fish (albeit small) in Moss Landing the day we shadowed CSF fisherman Calder Deyerly. This fisherman spoke eloquently and passionately about the craft of fishing, proudly passed down from one generation to the next, and the care he clearly had for his home coast and the marine ecosystems on which his livelihood depended. These are the kinds of fisherman we need and want, engaging in a form of sustainable fishing on a reasonable scale. Nevertheless, when I saw firsthand the writhing fish, it ignited a sickening feeling—these were clearly sentient, feeling creatures that were dying, or had died.

As with most food purchased in a modern grocery store there are few opportunities for (and little interest in) experiencing firsthand the difficult (and painful) lives of farm animals. Even fewer opportunities exist to see the harvesting of wild caught fish. In ways similar to factory farming, much or most of this happens hidden or masked from the consumers who ultimately buy and eat these fish.

The level of "bycatch," or the unintentional killing or harming associated with fishing is yet another reason for ethical concern. By some estimates the extent of global bycatch may reach 10.3 million tonnes per year.[8]

In early 2017, hundreds of common dolphins began washing ashore in the UK, France and Ireland, the likely result of offshore fishing trawlers. The number of dolphins killed has been estimated at more than 3000, a dramatic demonstration of the harm industrial fishing especially inflicts on marine mammals. This is just one episode that illustrates the huge problem of secondary death and suffering that happens in the regular course of commercial fishing. And while progress has been made to reduce bycatch—for instance the installation of turtle excluder devices—the problem and impact remain large and mostly unmitigated.

In contrast to a system based on harvesting fish, filter-feeding species such as mussels and oysters can be cultivated without this immense suffering. Singer believes the sentience line is somewhere between an oyster and a lobster or a shrimp, a speculation contained in *Animal Liberation*, and now being validated, Singer believes, by science. Their production and harvesting do not entail the same degree of environmental damage and, indeed, are actually ecologically restorative (e.g. filtering and cleansing water).

"So yes, if you want to eat animal products, oysters, clams and mussels...scallops are the best possible, most sustainable and least suffering animal products you can eat," Singer says.

There was a time when shellfish fed many Americans. Today, major shellfish ecosystems contain and produce only a fraction of what they did at their peak (Gulf of Mexico 10 % of peak, Chesapeake Bay only 1 %).[9] That could change and it is not inconceivable that oyster fisheries could be rebuilt, but it will require a major effort. There is little doubt that there is major growth potential for farm shellfish—oyster, clams, mussels—and a considerable amount of this might happen near cities. Efforts to engage urban residents directly in the raising of oysters—for instance through the Billion Oyster Project in New York, described in a later chapter—will further raise the visibility of the benefits of bivalves, even though water quality remains too poor to allow their human consumption. Bivalve production can be expanded in many parts of the country (and world) with positive environmental benefits (in contrast to the negatives associated with the raising of carnivores, that Pauly so well critiques). And we might envision a time when water quality is so improved in city harbors that oyster and shellfish could be raised for urban consumption (Fig. 3.6).

Cultivating seaweed is another promising option, and one that would meet Singer's sentience test. Like oysters and other shellfish, seaweed contributes positively to the marine environment and does not require input. A recent story in the *Washington Post* sings its praises:

> It needs no fertilizing, no weeding, no watering, and it has very few enemies in the form of pests or disease. It gets all its needs from the environment around it and, under optimal conditions, can grow almost six inches a day. It's healthful for people, and it actually leaves the environment better than it finds it.[10]

There are now a number of seaweed or kelp farms in operation, in the USA and around the world.[11]

Seaweed is not without its challenges. While nutritious, its high iodine content raises concerns about the potential health impact of eating too

**Fig 3.6** Oyster aquaculture is one promising and more sustainable option for producing food from the sea. In New York Harbor, the Billion Oyster Project is attempting to use oysters to educate residents on and connect residents to the harbor ecosystem, and is directly engaging high school students in this program. Image credit: New York Harbor School

much seaweed. And then for some there is the issue of taste (or lack thereof), and the important step of cultivating demand for this less-than-savory food product. But there is considerable promise in cultivating the sea in these ways, and through kelp and sea vegetables, and oyster and shellfish aquaculture, satisfying global protein needs in more sustainable and humane ways.

## Some Concluding Remarks

As many global fisheries crash and as our current largely industrial approach to seafood harvesting is proving unsustainable, Blue Biophilic Cities have the chance to lead the way to explore new ideas and avenues. There are few areas where the connection between urban consumption and marine health is clearer or more ripe for re-thinking. The issues are complex but some of the more promising avenues have been identified here. These include: greater support (financial and otherwise) for more community-based, small scale fishing (such as CSFs), and support for an emerging "slow fish" movement that parallels slow food and local food movements; utilizing the fish we harvest more efficiently and, as much as we can, eating further down the food chain; support for stronger fisheries management regimes, and the need to dramatically expand the extent of protected marine areas; and finding ways to shift consumption from wild-caught fish through more sustainable and humane forms of aquaculture. Blue biophilic cities give priority to ensuring healthy oceans but recognize that in many parts of the world wild-caught fish will remain an important source of protein. I have introduced the controversial topic of pain and suffering of marine organisms and, while the question remains open for discussion, the ethical underpinnings of Blue biophilic cities lend support to humane modes of aquaculture (i.e. oysters, clams, seaweed and other sea vegetables). Blue biophilic cities, moreover, support expanding protected marine areas, with special attention to those near to cities where urbanites can explore and learn about the marine world, but also cultivate an ethic of wonder and caring that can (it is hoped) extend to marine conservation in far-away locales.

## Notes

1. These include so-called TURFS or Territorial Use Rights for Fishing (TURF) Programs, see: http://fisherysolutionscenter.edf.org/catch-share-basics/turfs
2. Oregon State University, "International Science Team: Marine Reserves Can Help Mitigate Climate Change," June 5, 2017.

3. There are many ways that protected marine areas can enhance ocean resilience, for instance helping to address the increasing ocean water acidity: "coastal wetlands—including mangroves, seagrasses and salt marshes—have demonstrated a capacity for reducing local carbon dioxide concentrations because many contain plants with high rates of photosynthesis." Oregon State University, 2017, p. 2.

4. For more information see http://www.seafoodwatch.org/; there is also a discussion in Beatley, *Blue Urbanism*, Island Press, 2014a.

5. The authors also estimate that some 90 % of the "capacity enhancing subsidies" go to large scale fishing, supporting overfishing and marine degradation, "Conclusions indicate that taxpayer's money should be used to support sustainable fishing practices and in turn ocean conservation, and not to foster the degradation of marine ecosystems, often a result of capacity-enhancing subsidies. Reducing capacity-enhancing subsidies will have minimal negative effects on SSF [small-scale fisheries] communities since they receive very little of these subsidies to begin with. Instead, it will help to correct the existing inequality, enhance SSF economic viability, and promote global fisheries sustainability." Schuhbauer, Chuenpagdee, Cheung, Greer, and Sumalia, "How Subsidies Affect the Economic Viability of Small-Scale Fisheries," *Marine Policy*, 82, 114, August 2017.

6. Peter Singer, *Animal Liberation: A New Ethics for Our Treatment of Animals*, HarperCollins, 1975.

7. Interview with Peter Singer, April 4, 2017.

8. "How is Seafood Caught? A Look at Fishing Gear Types in Canada," found at: http://www.oceana.ca/en/blog/how-seafood-caught-look-fishing-gear-types-canada

9. Monica Jain, "Oysters Built the East Coast. Entrepreneurs are Rebuilding the Oysters," found at: http://voices.nationalgeographic.com/2017/04/11/oysters/

10. Tamar Haspel, "Seaweed is Easy to Grow, Sustainable and Nutritious. But it will never be Kale," *The Washington Post*, October 27, 2015.

11. One such company is Ocean Approved, located on the Maine coast. They advertise "Fresh, Frozen Kelp from Maine," with products that include kelp cubes and seaweed salad. See: http://www.oceanapproved.com/

# CHAPTER 4

# Making the Marine World Visible: Fostering Emotional Connections to the Sea

**Abstract** The marine world is, to many, remote and exotic. For city residents to fully embrace the wonder and beauty of the ocean world, and to actively work on its behalf, it will require emotional connection and caring. There are many different ways to do this and several of the more compelling and creative are described here: using social media to foster a sense of fascination and concern for the great white shark; taking children into the water and challenging them to find, look, touch and learn about the nature there; sending real-time video images from underwater divers to the surface; developing new long term institutions, such as a New York Harbor School and the Billion Oyster Project, to educate and engage residents of all ages. There are now compelling models that other cities can follow to foster this deep sense of emotional connection and caring for the marine realm.

Part of the difficulty of connecting urbanites with the marine world is that even though it may be nearby, it is often visually and emotionally distant. Our terrestrial world and lives make it hard to see and appreciate what lies beneath waves and water. Scuba diving and snorkeling offer a chance to see those worlds, but it seems unlikely that many residents will be able to do that.

© The Author(s) 2018
T. Beatley, *Blue Biophilic Cities*, Cities and the Global Politics of the
Environment, https://doi.org/10.1007/978-3-319-67955-6_4

There is as sense now that we must take full advantage of all the many ways, some conventional, some newer and more innovative, to reach and teach urban populations.

Visiting an aquarium might represent one opportunity for learning and for emotional connections, and we are now seeing a change in the ways in which these facilities are understanding their roles on the blue planet (for instance the new design for expansion of the National Aquarium in Baltimore). In San Francisco, we visited and filmed the Aquarium of the Bay. As aquaria go, it is a small operation, but it plays a big educational role in the region and it sits in the heart of the city, adjacent to Pier 39. Unlike many other aquaria, this one has only marine organisms that are native to and found in the ecosystems of San Francisco Bay. One of the aquarium's campaigns is cleverly titled "The Sharks of Alcatraz." It's meant to highlight the fact that there are indeed many species of sharks that can be found in the bay. The Alcatraz reference pertains to the story told that inmates of this island prison don't attempt to escape because if you do you will be eaten by sharks. Not really true, it was a convenient fear to stoke, and a story that provides a chance to educate about the remarkable shark life found close by.

Organizations such as the Georgia Strait Alliance in Vancouver has worked in several creative ways to build new appreciation for the ways that oceans might be under threat, and specifically addressing how urban lifestyles and recreational choices can have an impact. It has developed a series of online pledges aimed at both educating and building commitments to reducing pesticide and herbicide use on one's lawn, for example, and greener, more sustainable boating practices.

One of the most effective efforts at connecting urban residents to nearby water has been the use of "drift cards" as a way to simulate the potential impacts of oil spills. A joint initiative with the more science-based organization Raincoast, the cards are designed to float and are made of natural materials (plywood and nontoxic paint), intended to simulate an oil spill. The cards are tossed into the water from chosen locations, usually from a boat (often as part of an event organized with a local school), with the intent of understanding how far and fast the cards would travel. Each card carries a unique number and once citizens find them they are encouraged to go online to register where they found them, providing some highly useful information about how far and how quickly oil might move in the case of a real spill. It is an eye-opening exercise and the findings so far suggest that an oil spill would likely have a much wider geographical impact than previously thought, and the impact would be fast (Fig. 4.1).

**Fig. 4.1** The release of "drift cards" shown here, is a way to engage the public in studying how quickly and extensively potential oil spills from tankers might travel through the Strait of Georgia, near Vancouver, British Columbia. Image credit: Andrea Reimer, Georgia Strait Alliance

This is not an academic exercise. The Georgia Strait Alliance and other ocean conservation organizations are actively opposing a new proposed oil pipeline that would bring oil from the tar sands of Alberta province to be put on tankers in Barnaby (near Vancouver), which then travel through the Straits of Georgia.

The use of drift cards helps to educate about the oil threat and to foster a sense of caring about the ocean. According to Christianne Wilhelmson, the drift card program "has been incredibly powerful as a [public outreach] tool." Impacts that are abstract—a future oil spill—are made tangible and real in this way.

## EXCHANGING TWEETS WITH A GREAT WHITE SHARK

We need to further cultivate our "blue heartstrings," for lack of a better way to say it. There are natural times when we can emphasize empathy and caring for marine life and reject fear and separateness. The 1970s film *Jaws* did much damage to our collective psyche about sharks, teaching us that they are intrinsically dangerous and to be feared. That seems to be changing gradually, and it is heartening to see instances of herculean efforts on the parts of beachgoers to save stranded great white sharks.

Similarly, our new digital technologies and lives are helping us make connections and grow some of these much needed blue heartstrings.

The nonprofit Ocearch, for instance, has been tagging sharks and providing real-time information online about their whereabouts. These tagged sharks include a great white shark named Mary Lee, who has a Twitter account with more than 100,000 followers. Mary Lee tweets (in fact it was recently discovered that the real author of her tweets is a reporter for the Raleigh *News and Observer*) and her followers tweet her back, sending a variety of personal messages, from wishing her a happy mother's day, to encouraging her to return soon to their state and region. Along the way, followers seem to be learning about the shark—there are images, there is information about weight and distance traveled, and in the end perhaps a sense of something familiar; a digital friendship that helps to overcome the remoteness, the strong sense of "otherness" that a creature like a great white shark engenders.

I spoke to Ocearch founder Chris Fischer recently who discussed the ways in which shark tagging has helped to overcome the perhaps understandable disconnect (or aversion) that many feel towards sharks.[1] Fischer and his Ocearch crew have tagged some 300 sharks, including about 80 great whites (Mary Lee is in fact named after Chris's mom). The trips are usually collaborations with marine scientists from Woods Hole and the Mote Marine Laboratory. Fischer points to the value of this more public-inclusive mode of science and notes the many biological insights the trips and the tagging have generated.

Fischer argues that most of the fear of sharks stems from a fear of the unknown. "The only time we heard about a shark was when something bad happened and now we're talking about could Mary Lee be pregnant? Where is she giving birth? Where is the mating site?" Facts and curiosity and wonder replace fear. The secret seems to be to find ways to engage the public, to get their attention, to interest them in caring about these creatures.

"And we're having thousands of ongoing conversations throughout every day of the year," says Fischer. "Instead of just the odd shark attack story really driving how people feel... and so we're replacing kind of this fear of the unknown with really kind of the first facts and information that people can see and be a part of and that's allowed us to engage them in not only solving the problem of where they're mating, giving birth and migrating, but also to help them then understand why sharks are important."

This more inclusive model of science gives the public different avenues for contact and connection, "by allowing people to find their way into the project whether it's communicating with the shark on Twitter or tracking a shark on the tracker and then Tweeting or Facebooking a scientist with a question and connecting all these dots for people in real time, in the now." The Ocearch Facebook page now has more than 440,000 likes, so it's content and photos are clearly being seen (Fig. 4.2).

These modern digital tools are also proving to be helpful in the classroom, where elementary school students are following sharks in real time, learning about their biology, writing in journals about this, and generally replacing fear with fascination. Fischer tells me they have been working with a dozen schools to integrate a K-12 educational curriculum around the sharks, and to use this information in teaching other subjects from math to physics.

**Fig. 4.2** The nonprofit Ocearch has been tagging and tracking great white sharks and, through social media, helping to educate about and build new emotional connections to these majestic marine animals. Image credit: Ocearch

There is the impressive example of the first-grade class of Highlands Elementary School in New Jersey who have been keeping track of Mary Lee, and have even made a 16-foot paper replica of her that adorns the front wall of their classroom.[2]

These students are quite fond of Mary Lee, according to their teacher, Colleen Acerra, capturing their thoughts in a journal. "They love her," said Acerra, who was quoted in the *Asbury Park Press*. "They love tracking her. They love learning all about her and they're wondering if she's pregnant...Some people think she's pregnant and some people think she's just following the tuna run up and down the coast."

Fischer sees real change happening in the way sharks are being perceived and points to a recent episode where Cape Cod beach goers worked frantically to save a 14-foot great white shark. It was a remarkable example of collective compassion on Wellfleet Beach in Cape Cod, where a group of 100 human would-be rescuers carried buckets of water and dug a channel in an attempt to pull the fish back to open water.

Local reporter Alison Pohle reported on the scene:

> After people dug a trench around the shark, someone tied a rope to its tail. Another person then swam to a boat offshore with the other end of the rope. As the boat traveled away from the shore, the crowd dragged the shark back to the water, but it was too late.[3]

While unable in the end to save the shark, the effort was itself remarkable. The tide seems to be turning in our view of sharks. The image of people digging in the sand, passing buckets of water, pulling together on a rope in an effort to get a shark back to open water—what turned out to be a futile effort—is impressive to see.[4] And while it may not be the direct result of Mary Lee tweets, these social media connections are undoubtedly helpful.

Such efforts can lead to real and significant scientific insights, to be sure, and can result in more effective management and protection. Tagging and tracking of marine organisms, including sharks, has also happened through the Global Tagging of Pelagic Predators. Barbara Block of Stanford has been a leading force behind this tagging effort of sharks, but also bluefin tuna, elephant seals and California sea lions, among other species. Some species such as Pacific bluefin tuna are doing quite poorly, with populations estimated at 5 % of what they had been before extensive commercial fishing. How to protect and manage this species is a real challenge (as Block and her colleagues well know) and is the impetus for organizing a recent Bluefin Futures Symposium held at the Monterey Bay Aquarium. Tracking tuna has led to an understanding of where essential feeding grounds lie, including the area off the northwest coast of the USA where nutrient uplifting happens every spring—what Block has called our "blue Serengeti" (a compelling illustration of the power of language to help us connect with and understand the importance of this essential piece of seascape).

Can tracking this species ignite a level of engagement and concern around tuna and perhaps create the political space and cover for the tough management and conservation decisions necessary to ensure that Pacific bluefin tuna doesn't, as Block says, "go the way of the cod"? It is hard to know, but the new ways that technology may allow us to "wire the ocean" (through a network of wi-fi buoys and wave gliders—the latter being surfboard-like autonomous floating structures capable of collecting a variety of kinds of data from oceans).

Of course, there are many other creative ways that our modern digital technologies can foster nature connections. Our iPhones and tablets provide an almost unlimited opportunity to record the natural world around us and to effortlessly share these images, observations and experiences with friends and family. We are able prod, induce and incite with our Twitter posts, and Twitter and other social-media campaigns have proven to be effective methods of encouraging more natureful lives.

### CULTIVATING A MARINE ETHOS THROUGH EXPERIENCE AND ADVENTURE

How else do we develop a sense of connection and empathy? Another way to make this magic visceral and real is through direct experiences in or near the water. In blue cities we have filmed these kinds of experiences.

One such remarkable effort can be seen in the work of the Marjory Stoneman Douglas Biscayne Nature Center, a part of Crandon Park. With the Miami Beach Skyline seen in the distance, the center puts on a variety of impressive programs to educate in a very hands-on fashion about marine biology and marine conservation. At the center we filmed a visit hosted by Theodora Long, the center's long-serving director.

We watched as the center hosted a visit for some 100 fifth-graders from the inland community of Homestead, a community hard-hit by hurricane Andrew. The students initially sat and listened to one naturalist explain what they intended to do and what they might see and learn. Then we made our way out to the water for a "Seagrass Adventure."

The namesake of center, Marjory Stoneman Douglas, was an inspiration: a conservation hero in Florida history. Surviving until the age of 108, she was an active conservation advocate until the very end of her life. According to Long, she was intimately involved in the establishment of the center. She is most famous for her book on the Everglades, *River of Grass*, published in 1947.[5] It was the result of a writing contract—a request to contribute to a book series about rivers. The editor of this series first proposed that Douglas write about the Miami River, what Douglas thought to be a relatively small and unimportant river (and thus writing task). The Everglades, she thought, would be a much more appropriate subject, though before her book it was not generally viewed as a "river," but indeed it was on a grand scale. The book helped to change this understanding as well as to foster greater appreciation for a kind of natural environment that perhaps did not fit the more classical view of a picturesque landscape worthy of conservation (Figs. 4.3 and 4.4).

The center is a nonprofit organization in partnership with two other entities: the Miami Dade County Public Schools, and the Dade County Parks and Recreation Department.

On this beautiful Florida January day, with the skyline of Miami Beach in the distance, an energetic, excited group of fifth graders set off for a marine adventure. The almost 100 students from Homestead, an inland city to the west, were equipped with life vests and large nets for scooping the sandy bottom of the sea. They set off into thigh-high water in groups of around eight or ten. Each of the six or seven groups was accompanied by a naturalist from the center. There were also a few parents, but they seemed more hesitant about entering the water, one of them asking me sheepishly about whether there were any Portuguese man 'o wars around (there were, something the children had been briefed on but not too worried about). Once in the water the children, working in pairs, went to work and in no time there were screams of delight about what they were finding—a variety of small fish (e.g. trunkfish, toadfish), shrimp and sponges.

**Figs. 4.3 and 4.4**    One of the most effective ways to engage children in the marine realm is to take them into the ocean. Shown here are images from the so-called Seagrass Adventure, where fifth-graders are lead into the water to see what kinds of marine life they can find. This is one of the most popular programs of the Marjory Stoneman Douglas Biscayne Nature Center, Miami. Image credit: Tim Beatley

The children would bring their nets to the naturalist who would, with a smaller aquarium style net transfer the contents to a couple of plastic buckets floating (cleverly) in the center of two round life preservers. For me, one of the most dramatic early finds was what at first looked like a spiky tennis ball. I thought it was a species of sea urchin, until it was placed in water in the bucket, where it proceeded to miraculously transform back into a balloonfish. There were other dramatic finds—a queen conch, with its beautiful pink shell, a Filefish and blue crabs, sea hares and a spectacular bristleworm.

After more than an hour in the water the children came back to the shore to look more closely at and further discuss what they had found. The naturalists were terrific at simultaneously stoking the imaginations of these children, providing lots of additional information and natural history about the marine life, and managing their youthful impulses and exuberances (speaking all at once or moving around in distracting ways).

There was a lot of learning that happened on that day, reinforced by the visceral, hands-on aspect of the experience. The pupils learned that a tiny pea-sized fish would likely grow into a trunkfish, and that a mantis shrimp is in fact the fastest animal in the world (move over peregrine falcons) because of the rapid strikes of its claw. But I think the larger messages, more implicit, and more powerful were that a remarkable diversity of life lives just beyond our sight; that we can barely imagine what we might see in that net when it comes to the surface; that the sea is so magical and so different from anything else we know in our more land-based lives. The balloonfish certainly showed me that on that special day on Key Biscayne.

Filming this was fun and a privilege. We went from group to group, asking the pupils what they had found, and peering into the buckets to see for ourselves what had been collected (and to try to capture some of that on film).

There was also an indoor classroom element to this visit for the students. There are several classrooms at the center but one especially intriguing room, organized into a series of marine science "stations," each with a large aquarium in front of it. The children worked through a series of learning modules, answering questions by looking in the tanks or by accessing the other hands-on materials around them. A whistle blew periodically and the pupils shifted from one station to the next, gradually making their way around the room.

There are three full-time Miami-Dade County teachers who are assigned here permanently, employees of the school system (one of the main partners of the center, along with the Parks and Recreation Department).

Interviewing the center's director, Theodora Long, and program coordinator Sandra St. Hilaire, further helped to put this educational effort into a wider context. We heard the history of the center, going back to a time when they ran educational programs out of the back of a hotdog stand. Since 1969, Long tells me, they have reached more than a million people. This is not bad in a metropolitan area of 2.6 million.

Standing at Crandon Park, not far from the center, the skyline of the south end of Miami Beach can be seen in the distance. The prognosis for this city is not great, considering the rise in sea levels it will likely experience by the end of the century. But the city is bullish and resistant to the idea that it is ever going away or going to beaten by sea level rise, as we saw in the previous chapter. Proximity to this wondrous marine world is certainly one reason that helps to explain the deep draw of this coastal setting, though likely even well embedded adults have not had the full immersion marine experience provided by Long and her passionate staff.

## PIERING INTO THE NIGHT

We will need to look for more creative ideas to stimulate the imaginations of urban residents about what lies around and below. Luckily we have the technology to do this today. One innovative effort in the USA's northwest is called "Pier Into The Night." Organized by the nonprofit Harbor WildWatch, the idea is relatively simple—with the help of volunteer divers, underwater video is streamed in real time to a screen on a public pier. The divers carry a relatively inexpensive camera mounted with lights. They swim underwater looking for and pointing out marine life. On the pier there is a HarborWatch naturalist providing commentary to the on-screen images.

It is yet another creative way of making the undersea marine world visible. We visited (and filmed) one of these Pier Into the Night events in January—they happen on the first Saturday of every month from October through to March, a time of the year when the waters of Puget Sound are less murky and it is easier to see marine life (Fig. 4.5).

We spoke with Lindsey Stover, executive director of Harbor WildWatch, before the event began. HarborWatch has been organizing these events since 2009, and Stover tells me that this is not the first or only place where they are doing some version of underwater video. There was a palpable

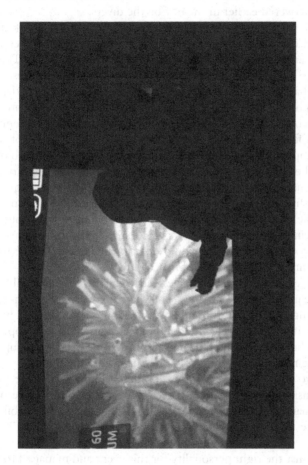

**Fig. 4.5** Pier Into the Night, is an event in Gig Harbor, Washington, where divers send images in real time to a screen on a public pier. Here families are mesmerized by what the divers are finding. Image credit: Tim Beatley

feeling of anticipation, of what might be discovered that evening. Stover describes the evening as an enticement to residents to experience "the mystery of what's below the surface."

It is "incredibly exciting" she says. "The kids love it. The big kids, all of us, the adults love it as well."[6] There are touch tanks, also, stocked with marine life discovered earlier in the day by the divers.

Everyone I tell about the Gig Harbor programs seems to think it is a great idea and I wonder why more places haven't tried this. As a reservoir of fascination and wonder, the marine environment has no equal. There is still so much to learn about the biology of marine organisms and these seem to live and behave in ways so completely foreign to what we see on land.

On the night we were filming there were about 90 residents huddled around the screen. It was relatively chilly, in the low 30s (Fahrenheit). Children and families mostly, I was impressed that there were not many sounds of complaining youngsters, anxious to leave for the warmth of car and home. The children seemed mesmerized by the images appearing on the screen. A naturalist from Harbor WildWatch, Stena Troyer, equipped with a hands-free microphone provided a steady stream of commentary about what the divers were finding and pointing to. There was a lot to see that evening, including sea cucumbers, decorator crabs and sea stars. The gloved finger or hand of a diver would occasionally appear on the screen directing attention to some piece of marine life.

The messages that evening were clear enough: there is remarkable and fascinating life just below the surface. "Anywhere there's space there is going to be stuff living," Stena told the audience. There was also trash and human debris to see, its own message. At one point a household iron appeared, allowing for a brief discussion of the need for beach and marine cleanups and the need to be more careful about what we throw into the water.

There were also fascinating stories about the biology of some of the marine organisms we were seeing. Troyer spoke with wonder in her voice about the remarkable things a sea cucumber can do, including its ability to spit out it intestines at a predator, if need be. "A pretty cool trick," Stena told us.

Stena had just the right personality for this event and managed to convey well a genuine sense of mystery and surprise at what was appearing on the screen. "The cool thing is that we never know what we're going to see. So, it is a different experience every time."

The assembled group got to see some remarkable forms of life, mostly hidden from view day to day. It was at least a glimpse into the hidden world of the marine realm and perhaps stirred further interest in children and adults alike. But the elusive giant Pacific octopus (or "GPO" as the divers referred to them affectionately) did not make an appearance. One had in fact been seen earlier in the day and so there was some hope for an appearance for a bigger audience later in the day. But no matter, the group was happy with what it had been able to see.

The Pier Into the Night is largely a volunteer effort. We spoke briefly with the two volunteer divers, Anthony and Allen, as they came out of the water at the end of the event. They were tired but excited to share their love of the underwater world with others. Much can be done it appears with a few volunteers and a few bits of relatively inexpensive technology.

## THE NEW YORK HARBOR SCHOOL AND THE POWER OF OYSTERS

One of the interesting programs under way in Baltimore, mentioned earlier in Chap. 2, was an initiative to engage the public in raising oysters. A collaboration between the Healthy Harbor Initiative and the Chesapeake Bay Foundation, some 150,000 young oysters (or spat as they're called) are being raised by schools and by corporate sponsors. These sponsors take on the task of helping to raise these spat until they are large enough to be transferred to oyster beds in Chesapeake Bay.

One innovation is that they are raised in hanging cages, not far away, but right there in the inner harbor. There are about ten sites where the hanging cages are found around the inner harbor. These so-called "oyster gardens" are at points accessible from the harbor's promenade, where schools and corporate volunteers can easily reach the cages to inspect them and clean them. Cleaning the cages must happen every month, Adam Lindquist tells me (Lindquist runs the program for the Healthy Harbor Initiative). When the cages are pulled to the surface, it attracts quite a bit of attention from passers-by. This Lindquist tells, is much of the point and provides an unusual chance to educate the wider public about the state of water quality and marine habitat. "All of my volunteers," Lindquist tells me, "are ambassadors for cleaning up the harbor, because they love to talk to the public." This is indeed a clever way to foster emotional connections with, and caring for, the aquatic realm of this city. I have been an enthusiastic admirer of the Baltimore effort and mentioned the thousands of oysters being raised there. When following one event, a resident of

New York City proudly declared, "Well, we are growing a billion oysters." In the months to follow I would learn much more about the impressive efforts to connect New Yorkers to their harbor through the growing of oysters, and the power of the oyster as both an educational tool and mode for connecting us to the marine world.

During our filming in New York City it was frequently mentioned that the city's historical motto had been the "City of Water." Though highly appropriate—it is a city with some 700 miles of shoreline—there has been an odd denial or disconnect with this profound watery context. That has been changing, as already noted, and New York Harbor is now commonly referred to as the city's sixth borough, and articles in outlets like the Smithsonian are declaring that New York is rediscovering its maritime spirit. These are all positive trends, but these shifts are not occurring by accident. The importance of doing many different things in that city is evident, from exploring new ideas for flood mitigation and sea level rise adaptation, to the new generation of waterfront parks like the Brooklyn Bridge Park, and these things are definitely transforming relationships with, and perceptions of, the harbor.

One of the most inspirational stories can be found in the New York Harbor School (I will return shortly to the Billion Oysters Project). We had the chance to interview Murray Fisher, founder of the school, on Governors Island where this unique New York high school is located. It is a spectacular setting, with views of the lower Manhattan skyline and considerable boat and ferry traffic in all directions.

Fisher told us the story of when he came up with the idea of a public high school centered on the harbor. He was working for the River Keepers organization, having moved to the big city from rural Virginia and with a background in farming. He began to realize that there few or no experiences for young people to learn about the harbor, or gather the kind of deep knowledge he had about the place in his position as river keeper. He had an epiphany. "This should be a school. This should be how we teach and learn. We should place young people directly in the ecosystems where they live." The light bulb went off, he told us.

The school he co-founded with several others is like no other. Part of the New York school system, it is considered a "career and technical education school," and students are able to graduate with technical credentials in one of six subjects taught: aquaculture; marine biology science; marine systems technology (including building and maintaining vessels); ocean engineering (including the operation of underwater ROVs, remotely operated vehicles); professional scuba diving; and vessel operations (that is driving and operating boats). Some 475 students travel to the high school from all five boroughs. Fisher explains that while one goal is for students to develop a skill set in one of these areas, another goal is to be accepted at the end to a four-year college or university. The high school reflects the philosophy that students should have maximum choice about their futures.

The array of skills and subjects taught is impressive indeed. We spent some time with two students who showed us around the aquaculture lab where they were working and learning. They walked us around and explained the tanks and magical looking vertical green-tinted cylinders (tanks of algae to feed the young oysters).

This high school is something to be proud of, and Fisher clearly is. It saw its first graduating class in 2007 and has just celebrated ten years of graduating enthusiastic, harbor-skilled students. Murray says he sees graduates working all over New York Harbor. "We feel the harbor school is helping to change the complexion of the industries that are so related to the marine environment." African–American students make up a high percentage of the high school's student body and historically few marine jobs and careers have been open to them.

"The basic big thesis behind the harbor school is that if you can create curriculum and job training skills and workforce development skills in high school that are directly related to restoring that ecosystem, then maybe we will have an opportunity to engage millions of youth in restoring the planet." It is a clever way to impart marketable skills but also a connection and care for the marine world (Fig. 4.6).

One of the higher visibility initiatives of the school has been the Billion Oyster Project (BOP), which has an audacious goal of planting and growing a billion oysters in the harbor. Fisher saw it as a way to bring together all of the students of the high school around a single big project. It was also a way to spread the message and curriculum to many other parts of the city beyond the high school. The project works with hundreds of students in a number of other schools, who learn about oysters and marine ecology and who help to plant and raise the oysters. The BOP curriculum is used in some 75 middle schools in New York. And now there are five "feeder" middle schools that have been designated, where students learn even more, and are more prepared to enter the Harbor School. So far some 20 million oysters have been planted.

Peter Malinowski, co-director of the BOP, spoke with us about how the BOP works in practice. All of the propagation and planting of the oysters

**Fig. 4.6** Billion Oyster Project engages high school students from the New York Harbor School in the raising, monitoring and studying of oysters placed in New York Harbor. Image credit: New York Harbor School

happens by the students at New York Harbor School. Those learning aquaculture raise the spat, those with a focus on marine biology monitor the health of the reefs, and so on. The BOP trains teachers at the middle schools and provides curricula. Students there learn math and science, Malinowski explained, "through the lens of oyster restoration." "Whatever you're learning in science class can be related back to the harbor. So rather than learning about prairie dogs and snakes, students learn about oyster toadfish and oysters, sea robins and sea horses, and all the animals that actually live here."[7]

Another key part of the project is the engagement of restaurants and their clientele. Some 60 New York City restaurants are participating in a program that collects and sends to the New York Harbor School their discarded oyster shells. After they are cured, they are used as oyster media and assembled into blocks that become the basis for new oyster beds. This is an important educational and community engagement function of the BOP. About a half million pounds of oyster shells were collected from these restaurants over the last year or so, but still more will be needed.

The planting of a billion oysters is a big goal and I wanted to know how much progress toward that goal had occurred thus far. Malinowski says that they have placed about 20 million new oysters into the harbor. Not close to the billion goal, but pretty impressive nonetheless. That is, until Malinowski suggests that there were likely trillions and trillions of oysters originally. By some estimates there were as many as 200,000 acres of oyster beds (a billion oysters would require, by comparison, about 200 acres).

The goal is not to plant anything near this original number, but rather to reach a point where the oysters are self-sustaining and where, as Malinowski says, they are able to reproduce on their own.

In the meantime, these new oysters, when placed, almost immediately enhance and improve marine habitat. But there is a larger goal, Fisher believes, and it is to build up a harbor-literate populous. "We've got to get the idea of nature and ecosystems back into New Yorker's minds," Fisher tells me. The BOP is a way to do that, a way "to grab people's attention."

Public education and ocean literacy must be at the core of Blue Biophilic Cities. Here we must circle back around again and mention the importance of the Waterfront Alliance and the 950 different organizations that make up this alliance that are actively engaged in marine education.

Summer programs are another opportunity. One especially good idea is the Harbor Camp, organized by several different groups, including the

Waterfront Alliance and the New York Harbor School. This program takes kids from disadvantaged neighborhoods in the city out on the water, exposing them to a side of the city they have not seen. Since its start, the program has reached some 20,000 kids, (and another 4000 are expected to participate in the summer of 2017). They learn about the harbor and its ecology, and they actually participate in a hands-on way with the sailing of vessels.

## SOME CONCLUDING THOUGHTS

Fisher describes New Yorkers as "shockingly disconnected" from that city's harbor and from its marine roots and history. There is a parallel story in almost every other coastal city. But there are many wonderful, creative ways to introduce urban populations to the marine nature all around them. This chapter has sampled and described several of the most compelling ideas, including sending video images of underwater marine life in real time to a waiting pier-bound audience; taking kids into the water to explore and see what they can find and identify; and engaging kids and the public in marine restoration efforts such as the Billion Oyster Project. There are things that can be done that will have a lasting impact, that will change, educate, entertain, deepen a sense of beyond one's narrow self, build larger communities and foster ultimately a sense of the wonder and majesty of the natural world. Maya Lin in her presentation at the Smithsonian's 2017 Earth Optimism Summit commented on the power of efforts such as the Billion Oyster Project to change the world: "If we can plant a billion oysters in New York Harbor we can save the planet."[8]

A few weeks after we filmed the wonderful Pier Into the Night event in Gig Harbor, we became aware of (indeed the nation heard about) the likely death of the matriarch of the southern resident orca population of the Salish Sea. Often referred to as J-2, she was also affectionately known as "Granny" for her longevity. Her death was a major blow on many levels and was felt acutely and deeply by the many human admirers that Granny had. In February of 2017 the Whale Museum in Friday Harbor, Washington, even hosted a potluck dinner to share stories about Granny.[9]

How we acknowledge and celebrate these marine lives, these creatures that have touched us, speaks volumes about our culture and also send signals to children and adults alike about the importance and inherent worth of the other living beings with whom we share the planet. It is hoped that some of the tools and techniques described herein will help to further strengthen these emotional human–marine bonds.

# NOTES

1. Phone interview with Chris Fischer, March, 2015.
2. See http://www.app.com/story/news/local/land-environment/enviroguy/2015/06/09/mary-lee-great-white-shark/28749931/
3. Alison Pohle, "Great White Shark Dies on Cape Cod Beach Despite Valiant Rescue Attempt," www.boston.com, September 7, 2015.
4. Watch the video here: https://www.bostonglobe.com/metro/2015/09/06/foot-great-white-washes-cape-cod-beach/tCZvIPuyPKsGAU2q4HBAKM/story.html
5. Marjory Stone Douglas, *The Everglades: River of Grass*, Rinehart and Company, 1947.
6. Personal interview and on-camera filming, Lindsey Stover, Gig Harbor, Washington, January, 2017.
7. Personal interview and on-site filming, Peter Malinowski, Governor's Harbor, New York City, May 9, 2017.
8. This quote was related to me by Murray Fisher, in an interview on Governors Island, May 8, 2017.
9. More about Granny and the Whale Museum's efforts here: https://whalemuseum.org/products/j-2-granny

# CHAPTER 5

# Rethinking the Blue–Urban Edge

**Abstract** Blue cities around the world are re-thinking their edges, as they must. New design and planning ideas include ensuring that the blue–urban residents have abundant physical and visual access to marine environments. New parks and shorelines are being designed to be more dynamic and multifunctional—projects such as the Dryline in New York City will at once enhance water connectivity, add public park space and urban nature, and provide for flood mitigation and resilience. There are new efforts, such as the Living Breakwaters project under way in Staten Island, to enhance ecology and connect residents to the water, while also enhancing the marine environment's resilience. New initiatives seek to integrate marine biodiversity into the design of new shoreline structures in cities such as Singapore and Seattle.

It is an ambitious plan to re-connect the city to the water, with new park space, a continuous promenade, re-built piers, places to launch kayaks and new street level connections to surrounding neighborhoods. Progress has already included completion of a first phase of a re-built Elliott Bay Seawall, designed to provide a new habitat for salmon (habitat benches and light-penetrating sidewalk panels), a set of unique textured surfaces that create spaces for marine invertebrates.

© The Author(s) 2018
T. Beatley, *Blue Biophilic Cities*, Cities and the Global Politics of the Environment, https://doi.org/10.1007/978-3-319-67955-6_5

In the larger framework of biophilic cities, we often advocate a "whole of city" approach, which recognizes the need for contact with nature at many levels and scales, and indeed at every scale. That is, it is important to bring nature inside of homes and offices (recognizing that we spend more than 90 % of our time inside), and we can do that in many creative ways (e.g. interior green walls, hanging plant systems, a variety of biophilic interior design ideas). But we also want to work to overcome the indoor/outdoor barriers by bringing the outside world in (e.g. by creative movable windows and doors, through greening terraces and balconies) and by imagining living building systems that add to the nature of the site and neighborhood (e.g. through exterior living walls). Every scale from "room to region", or "rooftop to region" (or bioregion), is a place where nature can be inserted, and we must protect and grow nature in all the urban spaces and places in between.

In coastal and marine cities this approach manifests in buildings and structures designed with natural features and elements throughout but which in turn flow through to and connect with the larger natural systems in which they are embedded, including rivers, harbors and waterfronts.

Given that we continue to spend the bulk of our day inside, their remains a need to bring nature into these working and living spaces. They should be designed to be as natureful as possible—with abundant living nature, natural light, water, and other biophilic features and, helping to overcome the indoor/outdoor barriers and thus making it easier to connect to outside nature.

There are an emerging set of good examples of waterfront or coastal buildings that achieve this. One is the 1 Hotel Brooklyn Bridge. A ten-storey, 194-room hotel, it emphasizes sustainability and connections with nature, and it displays the work of local artists.

A recent story on website *Curbed* critiques this hotel but notes the many ways its design incorporates the surrounding nature and its harbor setting:

'We had this running joke that it had to feel like it washed up on the shore,' says Drew Stuart, a partner at INC. Nothing here feels too precious; raw concrete pillars are visible throughout the hotel, and pallet-like wooden slabs are used as decor elements. Reclaimed wood from the Domino Sugar Factory is also used throughout.[1]

The lobby is bathed in natural light, and the wood and stone throughout make for a more biophilic feeling and design. There are a number of plants and other bits of living nature found here as well. Most dramatically is a living wall which is the focal point of the lobby and adjoining lounge. Designed by local landscape architecture firm Harrison Green, it snakes up and across the main interior wall, and is made of a series of soft-potted plants, connecting to a large wire grid mounted on the concrete. The effect is lovely (Figs. 5.1 and 5.2).

There are other biophilic and sustainability features in this hotel.[2] All the power needed for the building comes from wind energy, importantly, and of course there was a great effort to re-use building materials in its construction. There are also some clever smaller references to sustainability. I like the fact that all showers are equipped with five-minute hourglass timers. Even the room keys are sustainable—they contain flower seeds and are "plantable."

An example of a different kind of biophilic structure, providing similar biophilic connections and transitions between indoor and outdoor, is the Spaulding Rehabilitation Hospital in Boston. Completed in 2013, it was specifically sited on the harbor to take advantage of the therapeutic and healing power that water, and access to water, can provide. The site access, for instance, is Boston's Harborwalk, described as "a near-continuous, 43-mile linear park along Boston's shoreline" which "connects Boston's waterfront neighborhoods to Boston Harbor and each other."[3] The hospital has many biophilic and sustainable features. Described in a recent Urban Land Institute publication as a "resilient hospital," the first habitable floor is elevated 30 inches above the level of the 500-year flood and all mechanical systems have been placed on the roof.[4] Its system of backup generators will provide power for a number of days after a power outage, and operable windows (i.e. windows that can be opened) allow natural ventilation and allow occupants and patients to remain in the building.

Most interesting perhaps is the sense that the hospital is a healing building because of its proximity to the water. The windows in patient rooms are large and extend almost to the floor to allow patients in wheelchairs to enjoy the views. Visual access is important but so also is physical access to the water, and the possibility of launching a kayak or canoe from the site. Its flood- and wave-protecting landscaping, in the form of berms and swales, has also become part of the therapeutic program. Overall, the hospital demonstrates how it is possible to blend the resilient, sustainable and biophilic together, and effectively bridge the scales of building, site, neighborhood and city.

**Figs. 5.1 and 5.2**   1 Hotel Brooklyn Bridge is an example of a growing number of hotels that embrace sustainability. Most impressively, it incorporates biophilic design principles in many different ways, including through a prominent green wall in its lobby. Image credit: Tim Beatley

## SEEING AND EXPERIENCING BLUE CITIES IN SPECTACULAR NEW WAYS

Designing all buildings in blue cities so that they are resilient and bio-philic—office buildings, residential structures, university buildings and even hospitals—will be essential. But there are many other complementary ways that such cities can foster connections and connectedness to water, and there are some compelling examples of urban walks and coastal path-ways that at once provide a visitor or resident to fully immerse themselves, in a sense, in the watery realm.

Bridges can provide often quite spectacular opportunities to take in and experience the larger marine environment in cities such as New York, Sydney, San Francisco and Singapore. Walking across the Brooklyn Bridge has become a major attraction and features on many tourist bucket lists for good reason. It is a great way to experience the city, a chance to float above water and land, and to see the wider topography of the harbor (Fig. 5.3).

Sydney Harbor provides similar opportunities to see the water and under-stand the harbor setting from on high, in this case in the form of the city's iconic Harbor Bridge. Beginning in the late 1990s, visitors to Sydney have had the opportunity to climb the bridge and reach its apex. Tethered by a harness, climbing the bridge has become a major rite of passage in Sydney and a popular tourist attraction. Climbs are organized by a company appropriately called BridgeClimb, and each year it seems to add more fun elements. Sometimes climbers are met with live music when they reach the top. There are twilight and sunset climbs, and even a karaoke climb. The trip takes about 3.5 hours and is 1.75 kilometers in distance, and it requires traversing catwalks and climbing ladders. Some 3 million people have made the climb so far.[5] One of the FAQ's on the company's website is "Can I propose to my partner at the summit?", to which the answer is a decided yes. The visual vantage point offered on the beautiful harbor is said to be like no other (Fig. 5.4).

The Sydney Metropolitan area offers unparalleled access, visual and physical, to marine environments. Exemplary are the investments made by local councils and the state of New South Wales in coastal walkways that open up the marine edge to public visits. One dramatic example of this is the coastal walk from Coogee Beach to Bondi Beach (Figs. 5.5 and 5.6). It follows the rocky edge of this Hawkesbury Sandstone coastline, with spectacular views of the ocean along the way. The rugged coast appears in layers, with cliff-adapted flora and fauna, and birdlife throughout—willie wagtails, boisterous rainbow lorikeets, smaller more subtle birds though no less visually striking, such as the superb fairy wren. Wooden walking

**Fig. 5.3** The water views walking across the Brooklyn Bridge in New York City are spectacular. Image credit: Tim Beatley

**Fig. 5.4** The Sydney Harbor Bridge is a prominent landmark in the city, and it is now possible to climb to its top. Image credit: Tim Beatley

**Figs. 5.5 and 5.6**   One of the most spectacular and beautiful coastal walks is from Bondi Beach to Coogee Beach, in Sydney, Australia. It is popular with both residents and tourists. Image credit: Tim Beatley

structures hug the cliff, with stairs and switchbacks, and occasional opportunities to hike down to reach sheltered beaches and bays below. There are places to sit and rest along the way, with breathtaking perspectives. Many residents jog or walk the route daily, and there are occasional doorways providing direct access for coastal homeowners.

Another impressive stretch of this Eastern Beaches Coastal Walkway can be found south of Coogee Beach, a boardwalk through the Trenerry Reserve built in the 1990s. One especially memorable portion crosses one of the few remaining clifftop or rockshelf wetlands, with the sounds of frogs, such as the common eastern froglet, being audible. As the water runs over the rocky cliffs there are numerous mini-waterfalls, adding to the natural soundscape of this coastal hike.

Along the sea edges of the eastern beaches a series of small but biologically significant marine reserves have been established. These are spots that harbor remarkable marine biodiversity but are also important for their accessibility to urban residents, and their ability to offer snorkelers and divers a nearby view of the marine nature around them. One example is Gordon's Bay, a small marine protected area along the Coogee to Bondi walk. One innovation here is the Gordon's Bay Underwater Nature Trail, a 620 meter trail for divers and, on clear water days, snorkelers. The trail is maintained by the local scuba diving club and is easily accessible from a paved footpath. "The Trail can be compared to a bush walking track in the wilderness," declares a sign on the site, "only it is underwater" (Fig. 5.7).

To the north, near Manley Beach, can be found another example in the form of the Cabbage Tree Bay Aquatic Reserve. Not especially large, its value again derives in large part from its accessibility (Figs. 5.5 and 5.6).

Wellington, New Zealand, a partner city in the Biophilic Cities Network, offers a similar story of access to protected marine areas. In this case there is a marine life education center and a snorkel trail. The city has been rethinking its marine edges more comprehensively, working to develop "blue belts" that will complement its green belts.[6]

There are a variety of other ways urban residents can enjoy the watery edges and spaces, if cities make that possible. Extending our notion of greenspace to bluespace sends the signal that these are not empty and unimportant but worthy of visiting and enjoying. This can happen in many ways, including through boating and sailing, snorkeling and scuba diving, and increasingly swimming. Another positive outcome of improving water quality in ports and harbors in a number of cities, such as Copenhagen, is the establishment of public swimming areas, where previously it would have been unsafe and unimaginable.

**Fig. 5.7** Gordon's Bay Underwater Nature Trail, near Sydney, Australia, is very close to the shoreline and easily accessible by visitors and residents. Image credit: Tim Beatley

Swimming along the edges of cities has happened in some remarkable ways. The long history in Australia of establishing rockpools, or ocean pools, is one that exemplifies the possibilities of swimming in ways that connect to the larger ocean world.[7] In the Sydney suburb of Coogee (actually in the city of Randwick Council to be precise) there are some old examples of this impressive idea. North and south of Coogee Beach there are four rockpools open to the public, including one that dates back to the 1880s. These are essentially concrete pools crafted from the edges of the rugged cliff-lined coasts of eastern Australia.

They are bounded on the ocean side by a wall or two, depending on the shape of the pool, which allow ocean water and waves to enter, especially during high tides. Swimming in these ocean pools is a delight for many because it provides an experience of swimming in, or remarkably near, a wild ocean while being sheltered from its force.

The rockpools around Coogee are quite full of marine life. Fish swimming beside swimmers, and we are told that several octopi visit one. In the case of one another rockpool, I was told the story of the blue groper who arrived with the tide and stayed in the pool for around a month. A visit to the pool exposes one to one of the many signs erected by Randwick Council displaying the marine life likely to be encountered. Another beautiful and old rockpool can be found along the Manley Scenic Walkway, which connects Manley Beach to the small and sheltered Shelly Beach. Along another beautiful coastal walk, the ocean pool incorporates some beautiful sculptures with a marine theme (Fig. 5.8).

Along with the positive connection to marine life there are also be some small dangers associated with swimming in a rockpool. Entering Wylies pool, established in 1907, visitors are warned about stepping on sea urchins, which can be quite painful.

Re-discovering these shoreline and interstitial edge spaces around cities as places for swimming is something of a trend. Copenhagen has famously established a series of public swimming areas in the harbor, a result of major improvement in water quality. Paris recently instituted something similar along the Seine, and there are plans for the installation of "floating pools" in New York Harbor.[8]

Blue cities work hard to break through the barrier to water, to open up vistas and sightlines, and also to provide new opportunities for direct physical contact, whether by walking to the water's edge and touching the water, strolling along a beach, or providing opportunities for canoeing or kayaking.

New waterfront parks can provide this kind of access. A hallmark of New York's efforts at re-connecting to water can be seen in innovative

**Fig. 5.8** Australian rookpools provide an unusual opportunity for a more natureful swimming experience—one that blends the experience of a municipal pool with that of swimming in a more open ocean environment. Image credit: Tim Beatley

parks such the Hudson River Park and the Brooklyn Bridge Park. The Waterfront Alliance has been joining with a number of other organizations to expand physical access for boaters and kayakers. Especially along the shorelines of Queens, Brooklyn, Manhattan and the inner boroughs, it remains difficult to find spots to reach the water. Specifically, the Waterfront Alliance has joined forces with DockNYC and with the City of New York to expand the network of community eco docks in the region. These are floating docks that have been designed to better withstand the forces of a Hurricane Sandy type of event. The docks generally have two floating levels (lower level to accommodate kayaks and canoes), move up and down with the tide, and can serve as outdoor classrooms. They are even habitats themselves: "Fuzzy rope and oyster baskets hung from the sides and below will attract aquatic life and bring opportunities to study the marine life of New York's waterfront and harbor."[9]

## RESILIENCE IN BLUE–URBAN DESIGN

The experience of Hurricane Sandy has convinced many people of the need to design buildings that will be much more resilient and livable in the face of future storms. Sometimes referred to as "passive survivability" is the idea that structures should be habitable in the face of damaged power grids and other public infrastructure. Hurricane Katrina, which struck New Orleans in 2004, was a wake-up call and for many a realization that a quick national response and rescue from such events could not be guaranteed. In recent years, we have seen considerable attention being paid to designing new structures along or near the water limit the damage to them and ensure that they will remain livable in the days (and possibly weeks) following a storm event.

New high rise buildings in New York now include sufficient backup generators, powered by natural gas, to ensure continuation of some level of power and service. Generally there is a push to ensure that mechanical systems are elevated and placed at or above those floors likely to be impacted by floodwaters. The *New York Times* recently profiled one such new project, the American Copper Buildings. These are twin, connected towers (there is a skybridge joining the two), right on the East River, with distinctive architecture. The design includes five backup generators (located on the 48th floor) producing sufficient power to run elevators, a refrigerator and one outlet per apartment.[10] The walls of the building's lobby, which could likely see some floodwater in the future, are made from stone rather than wood. While flood resilience doesn't make it onto the developer's list of "top ten reasons why everyone wants to make an

American Copper home," number six is "waterfront property," and number three is "insane views" (they are spectacular water views).[11] The apartments have floor to ceiling windows and window walls on display on the project's sales website, and the water views are indeed remarkable.

In Baltimore we visited the site of the newly renovated Pier Hotel. A project of Kevin Plank, it has been elevated to the 500-year flood levels with a number of design features intended to respond to sea level rise. Kristen Baja, the city's chief resilience officer, described the structure's shift in window design—floor to ceiling windows were forbidden because of the chance that they might be subject to sea and wave action, and instead a flood-proof window system was required.

Organizations such as the Waterfront Alliance have also been actively encouraging different, more resilient designs for new waterfront projects. In particular, they have spearheaded the development of a new set of Waterfront Edge Design Guidelines (or WEDG), it represents a form of green building/site certification for waterfront development projects.

Even though they are voluntary, the guidelines are already having an impact and can be seen in the more resilient designs for re-developing the former Domino Sugar factory site, along the Brooklyn waterfront. Now under construction, the project's master plan (developed by the firm Field Operation) devotes space in front of the buildings to public waterfront park and public access, and the structures themselves are meant to be porous, "featuring large openings that allow light and air to penetrate through the site and into the neighborhood beyond."[12] The buildings have been set back, allowing the creation of a new quarter-mile waterfront park. Rather than blocking off the public from the water, the development seeks to actively connect to the surrounding neighborhood of Williamsburg.[13]

The Brooklyn Park Bridge, mentioned earlier, is the first waterfront park to gain WEDG certification. It is designed to better withstand future storms and flooding. As Roland Lewis described it during our interview there, resilience is a main goal, and this park proved its design during Hurricane Sandy: "The water came in, the salt edges did their job, let the water in where it could, and let the water back out after it receded, with minimal damage to the park infrastructure."

There are many things that cities can do to retain stormwater, and to address increases in flooding resulting from climate change. Cities such as Rotterdam have developed comprehensive strategies to better handle water, including design of urban fabric that helps it better retain water and in a sense act like an urban sponge. It has famously installed so-called "water plazas' designed to be community spaces most of the time but to collect and retain water when it rains. The first of these, Benthemplein, was completed in 2013. Designed by the firm De Urbaniste, the positioning of the different layers of spaces in the plaza is described in this way:[14]

> When its dry, the square is a feast for active youth to sport, play and linger. The first undeep basin is fit for everybody on wheels and whoever wants to watch them doing their thing. The second undeep basin will contain an island with a smooth "so you think you can dance" floor. The deep (third) basin is a true sports pit fit for football, volleyball and basketball, and is set up like a grand theatre to sit, see and be seen. On each entrance we create more intimate places to sit and linger. The planting plan emphasizes the beautiful existing trees. We plant high grasses and wild flowers surrounding the trees framed by a concrete border at seating height to offer many informal places to relax here.

A different version of this idea can be seen in the Western Australian capital of Perth. Completed in 2010, the Perth Urban Wetlands demonstrates that it is possible to convert a standard sterile urban water feature (energy intensive, heavily chlorinated) into something that can support native flora and fauna in the heart of the city. The re-design is the work Josh Byrne and his firm. Byrne is well known as a presenter for the national television show *Gardening Australia* and is the author of *The Green Gardener*. He speaks about the wetlands in a recent documentary film.[15] "The idea really was to provide an opportunity for reintroducing the types of plants and animals that were once common through this part of Perth before it was drained and became the city." From a biodiversity point of view the wetland has been quite successful, with flora and fauna thriving on the site. It is located in the heart of the city's cultural center, and the wetland sits on top of a museum storage facility. A lot of care went into plant and faunal selection. There are pygmy perch chosen to control mosquitos. Plants are thriving here, with varieties chosen in part to ensure they don't obscure views of the stage.

The project addresses several goals at once. The wetland is in the center of an amphitheater, with a stage in the background, so it helps to soften and draw people to these public spaces. There are now a variety of events that take place at the wetlands, including concerts, and sometimes a light show is projected onto the adjacent museum wall. The wetlands add an unusual dimension to the area, and Josh describes the setting of these performances as "quite magical."

It also adds a wonderful, cooling, biophilic element to the city's downtown area. There are stepping platforms that allow kids to walk into the wetland. It also has an important stormwater management function, collecting and filtering stormwater from surrounding buildings before it makes its way into the Swan River. "It's a great example of where a bit of inner urban biophilia and civic space can go hand in hand."[16]

## THE VISION OF A DYNAMIC SHORELINE

New York has been experimenting with other new ideas for re-imagining the shoreline as a more dynamic edge. The Big-U, more commonly now called the Dryline, is one important example of this new thinking. More formally called the East Side Coastal Resiliency Project, and with post-Sandy Federal Emergency Management Agency funding, this project envisions a 12 kilometer long "protective ribbon" encircling Lower Manhattan. Designed by the Danish architectural firm Bjarke Ingels Group (BIG), it is an important project in terms of how it re-imagines the shoreline and also how it integrates urban design and resiliency goals—it is planned as a series of berms, floodable park spaces and flood-resistant vegetation. It will be a place that residents will want to visit in fair weather, and a cityscape that will provide resilience in the face of a hurricane or coastal storm.

Kai-Uwe Bergmann, a partner at BIG, and the lead designer of the Dryline, spoke at the University of Virginia (UVA) about the inspiration for, and intentions behind, the innovative project. New York's Highline is a reference point not only in name but in its design program according to Bergmann, noting how the Highline uniquely re-imagines an infrastructure and imbues it with new social functions and purpose. Similarly we need a new kind of coastal infrastructure, a new approach to both protecting and contributing to the larger beauty and vitality of the city. He contrasts the Dryline with the sterile, single-function investments in floodwalls in New Orleans following Hurricane Katrina. "We wanted not to build a concrete wall but to build a flood protection system that adds to the quality of the city," Bergmann told the UVA audience:[17]

> It's a piece of urban furniture that also protects. It's a park that is there for everyone to enjoy 99.9% of the time, but then is also there to protect you against future storm events.[18]

Other cities are experimenting with similar ideas. Rotterdam has embraced the concept of tidal parks, which are designed to allow periodic flooding. The Brooklyn Bridge Park is designed in this way, as are other parks in floodplains or along rivers. The South Waterfront Park along the Monongahela River in Pittsburgh is yet another example (Fig. 5.9).

**Fig. 5.9** The "floating wetlands" have been placed in Baltimore's inner harbor. Adam Lindquist, shown here, runs the Healthy Harbor program, which is responsible for these floating wetlands. Image credit: Tim Beatley

Incorporating soft edges and healthy natural ecosystems along shore-lines will remain a preferred strategy where that is possible. There is now considerable research into the flood mitigation benefits of mangroves, for instance, and these, if they exist in or near city shorelines, need protect-ing.[19] Where possible, blue cities should pursue and use planning strate-gies and project designs that allow coastal wetlands and tidal marshes to move and migrate landward in response to sea level rise. Very often, how-ever, given the population density and extent of development present, that will be difficult. However, in many harbors, from Seattle to Sydney, there are a variety of coastal structures, from bulkheads to piers, where some degree of marine nature can exist.

## TOWARDS LIVING EDGES

These coastal edge designs and edge investments must also be seen as oppor-tunities to foster new life, to support the biodiversity of these places. As with the larger vision of biophilic cities, these too are spaces where we as humans can co-occupy and co-exist with many other forms of life, large and small.

Here there is also new thinking. I return to the work in Seattle around re-connecting that city with its shoreline. One of the elements is a highly unusual design for the new seawall. Part of a larger $410 million infra-structural investment, the seawall is intentionally designed to accommo-date human foot traffic on top, and through a series of light tiles, which allow light to penetrate to the wall below, the creation of what is essen-tially an "underwater corridor" habitat for salmon as they emerge from nearby rivers and make their way to the ocean. Much thinking has gone into the unusual design for the surfaces of the wall below. There are many different spaces and crevices, intended to support different kinds of marine organism, including "micro-algae."[20]

Heidi Hughes spoke of the unique ecological design elements of this seawall as we walked along and on top of the light-emitting glass tiles. The natural light making its way through is considerable, Hughes says, as is the marine life already taking hold below. The seawall panels are "designed with a variety of textures to encourage sea life. By raising the seabed with rock matrices, and putting this new seawall in, and the cantilevered light penetrating surface, we're creating a salmon migration corridor."

Hughes has a mockup of the seawall panels in the Waterfront Space, a kind of information center with a full scale model of the waterfront plan. "The whole design of the waterfront was made with the nearshore ecosystem in mind." Another interesting feature of the design, there are tidal markings which will let kayakers see where they are in the tidal cycle. Public art will also attempt to do some this connecting of visitors to the ebbs and flows of the water cycle.

Other cities, including Singapore and Sydney, have also been experimenting with new materials and designs for seawalls, revetments and other harbor structures so that they might also serve as habitats for marine organisms. In Sydney, efforts to "ecologize" these structures has led to the installation of "flowerpots" on seawalls, such as along the heritage wall at Sydney Royal Botanic Gardens. Designs have also included artificial rockpools and sediment-trapping elements that add marine habitat value.[21]

Singapore has also been experimenting with and testing different seawall surfaces. NParks researchers Nguyen, Tun and Chan describe an ongoing monitoring effort to test different forms of surface complexity and the extent to which marine organisms colonize them over time. Different surface configurations are being explored to see which combinations of pits, grooves and steps might attract the most marine biodiversity. While monitoring continues and conclusions have yet to be reached, the results so far are encouraging:

> Faunal diversity and abundance increased over time and, after several weeks, we recorded periwinkle and nerite snails, crabs, tube and fire worms, feather stars, and bead anemones. The performance of each tidal pool design and its complexity elements are also being monitored. The outcomes of this study are expected to provide a more comprehensive understanding of the combination of complexity treatments on species recruitment and biodiversity.[22]

The researchers see hope that such seawall structures will be designed to respond both to sea level rise and to the need to support marine biodiversity.

The re-design of urban aquaria offers the chance to soften edges and to re-introduce natural systems. One notable example is the ongoing re-design of the National Aquarium in Baltimore. Architect Jeanne Gang, and the Gang Studio, has spearheaded this effort, producing a *Strategic Master Plan* in 2015. Among other things, the plan imagines a new emphasis on educating about the Chesapeake Bay ecosystem, and generally shifting to a "new and more conservation-oriented visitor experience."[23] Most impressive is a new 37,000 square foot urban wetland that will serve as the centerpiece of the aquarium's outdoor space. Work has also begun to design a dolphin sanctuary, a place to humanely re-locate the aquarium's eight resident bottlenose dolphins.[24]

Currently the aquarium has installed and is monitoring several floating wetlands. The design and installation of floating wetlands has actually been a major project of the Healthy Harbor Initiative with considerable success. With design help from the firm BioHabitats, these floating wetlands have been assembled and installed with volunteer help. They utilize recycled plastic bottles for floatation. They have now been installed on 54 sites, covering some 2000 square feet of the inner harbor. They add much visually and contribute to a sense of a nature in an otherwise largely grey and builtup harbor environment, and the evidence is that they actually work to cleanse the water and add to the habitat. A number of harbor animals have been sighted using them, including blue heron and river otters, and a 2011 study concludes that they have been quickly colonized by marine organisms (and an estimated 0.5 million dark false mussels per platform). Adam Lindquist, who runs the Healthy Harbor Initiative, tells me that the floating wetlands have been popular: "Not only do people love the soft green vibe that they bring to an otherwise hard urban area, they also actually help to remove pollution as they grow, so the plants actually suck up the excess nutrients in the harbor."

Landscape architect Kate Orff and her firm SCAPE have been experimenting with a variety of living shoreline/harbor design ideas. Most famously is her idea of Oyster-Tecture, suggesting that we ought to design and install oyster beds in New York Harbor as both works of ecological restoration and as flood mitigation strategies. She developed this idea pre-Hurricane Sandy as part of the exhibition at the Metropolitan Museum of Art in New York, and it has received a widespread embrace.

The Living Breakwaters project extends this earlier work and scales it up. It involves installing two miles of breakwater structures along the south coast of Staten Island, seeded with oysters (planted through the Billion Oyster Project), in combination with onshore dunes. There is a heavy community engagement dimension as Orff is working with Staten Island schools to plant the oysters and monitor the marine life taking hold there. Part of what is unique about the scheme, Orff tells me, is the multifunctional aspect: "The project is aiming not just to solve one problem, but aiming to foster water-based culture again in Staten Island."[25] The breakwaters will engage kids, schools and adults, will provide protection and reduce risks from future flooding, and will help to regenerate harbor ecosystems ecologically.

For Orff, much of her design work is about overcoming the traditional Olmstedian dichotomies of passive versus non-passive landscapes. She sees the need to understand landscapes, and especially New York Harbor, as a living productive landscape, and one that humans can and must be actively engaged with.

The harbor, Orff argues, "can't just be a backdrop anymore." And she sees the need for a new kind of landscape architecture that sees water in a profoundly different way. "It's not just water as backdrop, as empty, as a space that is not land, [but] literally understanding that water as full of life."

## SOME CONCLUDING THOUGHTS

These are exciting times in blue biophilic cities as there are now many new ideas for re-thinking the shoreline edge in ways that will enhance habitat and human connection while also responding to flooding and sea level rise. The idea of a less fixed, profoundly dynamic edge seems better suited to the current world and to a more integrative land–sea vision of blue biophilic cities. The design toolbox is now much more expansive: it includes floodable parks and floating wetlands, habitat seawalls and rock-pools, and the use of multisurfaced and creviced facades that make room for the marine in cities. There are even more ambitious designs to be imagined in the future, and blue biophilic cities will be in a position to push the design and planning envelopes yet further in the years ahead.

## NOTES

1. Amy Plitt, "Tour 1 Hotels' New Sustainable NYC Hotel in Brooklyn Bridge Park," *Curbed*, found at: https://ny.curbed.com/2017/3/24/14902856/nyc-hotels-brooklyn-bridge-park-design, March 24, 2017.
2. E.g. see Zachary Weiss, "This New Brooklyn Hotel Runs Entirely on Wind Power," found at: http://observer.com/2017/02/1-hotel-brooklyn-bridge-review/, February 23, 2017.
3. "About the HarborWalk," found at: http://www.bostonharbornow.org/what-we-do/explore/harborwalk/
4. See Urban Land Institute, "Spaulding Rehabilitation Hospital," found at: http://returnsonresilience.uli.org/case/spaulding-rehabilitation-hospital/
5. For more information, see Sydney BridgeClimb: www.bridgeclimb.com

6. For more background on Wellington's efforts around blue belts, see Beatley, *Blue Urbanism: Connecting Cities and Oceans*, Washington, DC: Island Press, 2014.
7. The precise number of rockpools along the New South Wales coast is unclear but is likely between 50 and 100; e.g. "In Search of the South Coast's Best Ocean Pool," http://www.canberratimes.com.au/act-news/ canberra-life/in-search-of-the-south-coasts-best-ocean-pool- 20150303-13ttn0.html
8. Matt Hansen, "Swim in New York's East River? A Floating Pool Plan Envisions a Safe Way," *LA Times*, March 15, 2015, found at: http://www. latimes.com/nation/la-na-floating-pools-20150315-story.html
9. "First-of-its-kind $1.1 Million Eco Dock Opens On The South Brooklyn Waterfront," found at: https://www.nycgovparks.org/parks/american- veterans-memorial-pier/pressrelease/21190
10. David W. Dunlap, "Building to the Sky, With a Plan for High Waters," *New York Times*, January 26, 2017.
11. http://americancopper.nyc/
12. "Domino Sugar Refinery Master Plan," found at: www.shoparc.com/proj- ects/domino-sugar-refinery/
13. E.g. see Jules Gianakos, "Domino Sugar Factory Master Plan Development," *Arch Daily*, March 5, 2013.
14. Water Square Benthemplein, found at: http://www.urbanisten.nl/ wp/?portfolio=waterplein-benthemplein
15. Interview and site visit with Josh Byrne, July 5, 2017, See the film here: https://www.youtube.com/watch?v=LGJhcMdQyY8.
16. see Perth Urban Wetland film, https://www.youtube.com/watch?v= LGJhcMdQyY8.
17. Kai-Uwe Bergmann, lecture, April 3, 2017, University of Virginia, Charlottesville, VA.
18. Ibid.
19. E.g. see "Mangroves for Coastal Defence," found at: https://docs.google. com/document/d/1q__FT92WS-Bplh5Wwplao2e88Aou2BXpNz3A3KX_ GCo/edit
20. E.g. see Ken Christensen, "Seattle's New Seawall: Holding Back the Tide, Protecting Salmon," *crosscut.com*, May 18, 2017.
21. E.g. see K.A. Dafforn et al., *Guiding Principles for Marine Foreshore Developments: Report Prepared for Urban Growth NSW*, Sydney: UNSW, 2016, found at: http://thebayssydney.nsw.gov.au/assets/Document- Library/Precinct-Wide-Technical-Studies-Underway-2015-/2016- Guiding-Principles-for-Marine-Foreshore-Developments.pdf
22. Nguyen, Tun and Chan, "If You Build It, They Will Come: Modifying Coastal Structures for Habitat Enhancement," *The Nature of Cities Blog*,

October 5, 2016, found at: https://www.thenatureofcities.com/2016/10/ 05/if-you-build-it-they-will-come-modifying-coastal-structures-for-habitat-enhancement/

23. Gang Studio, *National Aquarium Strategic Master Plan*, found at: http:// studiogang.com/project/national-aquarium-blueprint

24. Edward Gunts, "National Aquarium in Baltimore to Build North America's First Sanctuary for "Retired" Dolphins," *Architects Newspaper*, June 14, 2016, found at: https://archpaper.com/2016/06/national-aquarium-dolphin-sanctuary-studio-gang/

25. Interview with and filming of Kate Orff, Battery Park, New York City, May 9, 2017.

CHAPTER 6

# *Just* Blue (and Biophilic) Cities

**Abstract** Social justice is a key goal in a blue biophilic city. Often there are significant social inequalities in terms of access to nature, including blue nature, and efforts such as the Blue Greenway in San Francisco and the new Waterfront Park in Newark seek to more fairly distribute these blue benefits. This chapter introduces the concept of equigenic blue: that investments in access to blue nature can help to overcome other forms of social and health disparity. There are other important aspects of social justice explored here, including the extent of ethical obligations to future generations and to other non-human forms of life.

Our blue cities of the future must also place social justice at the fore, and understand how both the benefits and the burdens of the blue realm are distributed across a city. Physical and visual access to blue medicine is often correlated with higher income, white neighborhoods, with lower income and ethnic minorities being less able to access and enjoy these marine urban values.

It is a common goal of biophilic cities to provide easy access to parks and nature for all, not just those with income and privilege. This is frequently expressed in terms of the percentage of residents within a 5–10 minute walk of a park or other green area. Experience shows that while

© The Author(s) 2018
T. Beatley, *Blue Biophilic Cities*, Cities and the Global Politics of the Environment, https://doi.org/10.1007/978-3-319-67955-6_6

many cities have made great strides towards achieving equal access to nature, profound inequalities remain in many cities.

Access, physical and visual, to water, and to oceanfront and waterfront locations, in blue cities represents a tremendous opportunity to address these inequalities. More economically distressed neighborhoods may lack conventional parks or greenery, but likely are not far away from a shoreline. In blue biophilic cities there is an emphasis on utilizing the marine and aquatic realm as a restorative asset to enhance the quality of life in these places.

More generally there is the goal that in our ambitious waterfront redevelopment visions, and in creating new public spaces to enjoy water, that we work hard to make these spaces attractive and accessible to all residents of the city. Efforts such as Seattle's New Waterfront have explicitly framed the vision as a "Waterfront for All," an encouraging sentiment and goal. In that city, creative experimentation through its Hot Spot pilot project has been looking for ways to entice residents to the waterfront, and to convey the message to all residents that there are things to experience and enjoy there. Through a variety of means including music, dance and even pop-up soccer, the work of groups like Friends of Waterfront Seattle aim to ensure that all residents feel welcome.

## EQUIGENIC BLUE

Cities such as San Francisco have taken steps to expand access to the blue realm. The Blue Greenway, for instance, seeks to expand and connect shoreline parks in some of the traditionally poorest parts of the city. The neighborhoods of Hunters Point and Bayview have been the site of considerable industrial contamination. A former shipyard and power plant are being cleaned up as part of this planning initiative. When this work is completed, residents will be able to reach the water more easily, with enhanced pedestrian and bicycle mobility as a result of the 13-mile Blue Greenway (Fig. 6.1).

Another compelling story about the power of water in helping cities and neighborhoods overcome disinvestment and economic and social hardships is that of Newark's Riverfront Park. Now with its third phase under way, it has provided new public access to the Passaic River, a waterway heavily contaminated by the chemical industry there. The scheme is spearheaded by the city's Newark Waterfront Revival Initiative, with the "aim to connect every Newarker to their river."[1] The river is slowly being

**Fig. 6.1** Through an initiative of the Baltimore Parks Department, kids from underserved neighborhoods learn all about kayaks, including how to sit in and steer them, and then they experience being on the water often for the first time. Image credit: Tim Beatley

cleaned, and along with it a new greenway and park along the river are being created, providing new river access and new public spaces for a variety of activities, from music to Zumba to arts festivals. There are new sports fields, a floating dock and a new boardwalk painted a distinctive orange. The impacts have already been considerable.

The new park and greenway, which will ultimately cover more than 20 acres, are the result of a collaboration between the city (and its initiative Newark Waterfront Revival) and the Trust for Public Land, with a variety of other partners. Of course there is an economic benefit to this new park, and also a health benefit to residents. In a city with a high concentration of underserved, the benefits of walking trails and parks are immense. One new health partnership can be seen in the Horizon Wellness Trail (financially supported by the Horizon Foundation), which will encourage healthy walking and is designed in a way that makes it easy to set and exceed daily walking targets.

These efforts at opening up access to the river go back to 2008, and the city has employed some creative ways to foster interest and community support. The planning department has organized hundreds of tours of the riverfront, some by boat and many on foot. "One of the things about the park is how good people are to each other when they're there," said a volunteer.[2] Waterfront spaces, and contact with water generally, are a vital urban salve, delivering some potential mental and physical health benefits, and as with other natureful settings they have the opportunity to enhance human connections and make us more generous, better human beings.

Likely at work here is what we might describe as "equigenic blue". Equigensis refers to the ways in which green spaces, access to nature and the quality of physical environments may help to reduce health inequalities. We would ideally like to see shifts in income, education and the many other ways in which health inequalities might be corrected, but the likelihood of this is low to nil, so the design and planning of environments become even more important. Research is emerging that shows that access to coastlines and water have the potential to be profoundly equigenic—to reduce the inequalities in health and wellbeing.[3] This is true for both San Francisco's Blue Greenway and Newark's Riverfront Park.

There are efforts in a number of cities to ensure that adequate opportunities exist in minority and underserved communities to enjoy the blue nature that may exist nearby. The City of Baltimore's program aimed at teaching kayaking and offering the chance to experience the inner harbor in a kayak is one such example. Its Healthy Harbor Initiative has also sought to engage underserved neighborhoods in other ways, including through its Alley Makeover program. We visited and filmed one alley where neighbors had come together to design and paint, with a remarkable result—a painted blueway—an uplifting piece of neighborhood art that has helped neighbors to meet each other and come together to improve their homes and watershed (Fig. 6.2).

Yet another important dimension of the "just blue" agenda is working hard to ensure that the many positive qualities of new waterfront parks and greenspaces do not unfairly result in the displacement and gentrification of the surrounding neighborhoods. An example frequently invoked is the High Line in New York City. An elevated former commercial rail line, running through the Chelsea neighborhood of Manhattan, it has been a

**Fig. 6.2**   An alleyway in an underserved neighborhood in Baltimore, Maryland, receives an "Alley Makeover." Image credit: photo by Adam Stab; alley art design by Adam Stab and Leanna Wetmore

wonderful and unique addition to the parks and nature of that city. It is an unusual green and biophilic ribbon of nature running above and through the city, providing spectacular views and impressive green spaces in which to relax and stroll. But these very impressive qualities have also resulted in massive amounts of new development being attracted to the area, raising the price of housing and displacing many residents. These unintended side-effects of otherwise exemplary biophilic projects have been described as a kind of "ecological gentrification."[4]

New parks in biophilic and blue cities must begin then to take a different approach and plan early to minimize these kinds of unintended consequences. One recent positive example can be seen in the proposed 11th Street Bridge Park, being planned for Washington, DC. Spanning the Anacostia River, this bridge has a strong community equity mission at its core—seeking to bridge the literal and income gulf that exists between the affluent west side of the river with the largely minority and lower-income neighborhoods east of the river. A unique design, the result of a design competition, the structure will provide an unusual pedestrian environment, complete with new public spaces, a cafe, an environmental education center, an urban farm, and even a large waterfall. (Fig. 6.3).

I spoke recently with Scott Kratz, who is coordinating this $45 million project under the auspices of the non-profit Building Bridges Across the River. From the beginning, an emphasis was given to listening to and consulting with residents. Some 200 meetings with community stakeholders were organized to determine park programming, and a later design competition for the bridge structured in a way that community input was central. Most impressively, the project is spearheading a comprehensive effort to ensure that the underserved east side of the river will benefit from the project and that the unintended consequences that are likely—higher-priced housing, displacement, and gentrification—are avoided or at least minimized. An impressive Equitable Development Plan has been prepared, and many of the proposed measures, from workforce development to forming a community land trust, are already under way.[5] Kratz tells me that ultimately he will judge the success of this piece of innovative urban infrastructure by how well these underserved neighborhoods will benefit economically and socially. "Who is this for?" has been a primary question from the beginning, and a fair and important lens for judging most blue, biophilic initiatives.

The question of fair access to the blue often raises fundamental questions about the appropriate line between public and private interest. As many cities such as New York, Boston and Seattle re-discover their waterfronts, the process of privatizing and monetizing these spaces leads to ongoing debates about fairness. Should coastal developers be permitted to profit immensely from the physical access and amenities that shorelines now create, and the spectacular views of water that are now cherished and rewarded in the marketplace. On one hand, we see the positive trend in cities such as New York especially of opening up new opportunities for waterfront access in places such as Brooklyn Bridge Park, but on the other

**Fig. 6.3** A rendering of what the 11th Street Bridge Park in Washington, DC, will look like when completed. Image credit: Courtesy OMA + OLIN

hand, the best locations to love and the best views will still be retained by those able to pay for them.

Boston is a city where similar re-development trends are occurring and a similar re-discovery and appreciation for its harbor has taken place. Some have been critical of the city's willingness to let private developers capture much of this positive coastal value, especially given that it has been the public money—investments by the public sector—in cleaning up the harbor that have made possible this waterfront renaissance. Organizations such as the Conservation Law Foundation (CLF) have taken legal action to ensure that the public's interest is adequately safeguarded in the review of waterfront development proposals. Peter Shelley, interim president of the CLF, makes the case succinctly in an interview in the *Boston Globe*: the public has spent billions to fix leaking sewer pipes and to clean up Boston Harbor, "And yet when development goes forward around the harbor, you're [the public] getting completely excluded ... That's an outrage."[6]

These discussions taking place in cities such as Boston about how people should enjoy or benefit from public investments in harbor and waterfront cleanups also introduces some important temporal questions. Blue cities favor the larger civic interest in accessing water, and doing so, like the cleanup of Boston Harbor, requires long-term commitments and thinking well beyond the narrow electoral cycle. Blue biophilic cities recognize that our moral communities—the people and things to which we have duties—extend beyond the narrow present. Indeed, many of our decisions, say commitments to protect cetaceans or to clean up ocean plastics, are motivated in part out of a duty to future generations.

Thinking longer term is not always easy for the human species, of course. But many of our toughest ocean challenges will require a deeper time perspective, and will require an ethical commitment to think carefully about and act on behalf of a healthy marine future. I recently had the chance to sit down with Long Now Foundation executive director Alexander Rose. The Long Now is an idea advanced by Stewart Brand, and laid out in a thin but highly influential book *The Clock of the Long Now*.[7] Rose talked at length about the concept of the Long Now and the utility and importance of longer term thinking, and where some of the Long Now's most iconic projects stand.

The core of the idea behind the Long Now is a need to break out of our incessant short-term decisions and decision making. Not that we need not make such decisions but that they should be framed and guided by a longer, deeper sense of time. The Long Now takes 10,000 years as an important marker—it has been about that amount of time since the human species developed agriculture. As Rose told me, "The idea behind the Long Now is both the last 10,000 years and the next 10,000 years .... We're placing ourselves in the middle of the story rather than at the end of a story."[8]

One of the key ideas here is that we must be ever cognizant of actions that will limit choices in the future. It is the principle of keeping options open, and avoiding actions that will close future society avenues. This is a key decision premise of the Long Now: avoid irreversible steps (think extinction of a species, destruction of a complex ecosystem such as a coral reef).

"Fundamentally, if you're making choices that limit future generations choices, you are making the wrong choices. If you're making choices that increase those choices, you're making the right choices." Pretty clearly much of what we are doing in the marine realm—rapidly overharvesting fish, filling oceans with garbage and plastics, and, of course, climate change—will have huge and lasting impacts and will limit future choices. Simply giving futurity a voice in the debate about how we use and treat our marine environments, near and far, would be a positive step, and many of the ideas advanced in this volume—expanding marine protected areas, limiting seafood harvests, re-wilding marine environments near to cities—could all be defended by reference to keeping options open and by a more cautious and sensible posture towards the future. And I would argue, moreover, that our efforts to design and create blue, biophilic cities will be option-expanding and more respectful of future generations.

There are a host of other challenging ethical issues and quandaries that emerge in the design and planning of blue and biophilic cities. Some have been surprises during our interviews with leaders in the marine world. In my interview with Daniel Pauly I was somewhat startled that the subject of slavery on the high seas figured so prominently in the conversation. One feature of the modern fishing industry is that ships are able to spend months, or even years, at sea, with labor that has been captured or tricked into service. This low wage labor makes up another significant subsidy to modern fishing and is a serious human rights concern. There is now more awareness of these issues as the popular press has given more attention to it, but I suspect that it is still a largely hidden dimension to which the current industrial seafood system is unjust.

At the heart of the concept of biophilia—indeed, what its intuitive meaning "love of nature" suggests—is a sense of caring about and concern for other forms of life. A biophilic city, as mentioned in Chap. 1, is understood as one of shared spaces, as a place where many different forms of life are found. For coastal and marine cities the challenge is to make that nearby life manifest and visible.

And increasingly we recognize the moral if not legal rights held by at least the larger, (perceived to be) more sentient and intelligent marine life that live and visit near to cities—whales, dolphins, marine mammals, for instance. Biophilia and the vision of biophilic cities broaden the moral community to which we have duties to include all life, all living things and all living systems. In these ways we (i.e. cities) have an ethical duty to respect and protect this marine world and all of its living inhabitants.

## Some Concluding Thoughts

In this chapter I have sought to at least begin the discussion about the social justice and social equity obligations of a vision of blue, biophilic cities. These are complex issues to be sure and they require community discussion and debate. But at the minimum every design and planning project or proposal should take its "justness" into account. Every effort should be made to ensure that we provide truly public access to the profoundly restorative and wondrous marine world, and that we employ connections and access to water as a proactive antidote to overcoming societal inequalities in health and opportunity. Finally, the blue biophilic city understands that justice requires recognition of the inherent moral worth of all marine life: we seek to enjoy it, to learn from it, to revel in it and to be inspired by it, but we also have a duty to protect and care for it.

## Notes

1. See Newark Riverfront Revival, found at: http://newarkriverfront.org/
2. Steve Strunsky, "Newark Breaks Ground on Riverfront Park Expansion," found at: http://www.nj.com/essex/index.ssf/2016/10/newark_riverfront_park_expansion_groundbreaking.html, October 5, 2016.
3. See Rich Mitchell, "What Is Equigenesis and How Might It Help Narrow Health Inequalities?" found at: https://cresh.org.uk/2013/11/08/what-is-equigenesis-and-how-might-it-help-narrow-health-inequalities/
    Mitchell hypothesizes several different ways in which equigenisis might work: "Equigenesis might work in two ways; levelling up or levelling down. An *equigenic* environment which levels up presumably supports the health of the less advantaged as much as, or perhaps more than, the more advantaged. An equigenic environment which levels down presumably limits the health of the more advantaged to a greater extent that the less advantaged.

Given our desire to improve population health overall, it would clearly be better to level up."

4. E.g. see Sarah Dooling, Ecological Gentrification: A Research Agenda Exploring Justice in the City, *International Journal of Urban and Regional Research*, 33, no. 3, 621–639, 2009.

5. See Equitable Development Plan, found at: http://bridgepark.org/sites/default/files/Resources/EDP%20Final%20-%20UPDATED.pdf

6. Thomas Farragher, "It's a Cleaner Harbor, with Fewer Spots to Enjoy It," *Boston Globe*, May 27, 2017, found at: https://www.bostonglobe.com/metro/2017/05/27/cleaner-harbor-with-fewer-spots-enjoy/ZzPRAks04xRcZjP7h2h8uM/story.html

7. Stewart Brand, *The Clock of The Long Now: Time and Responsibility*, Basic Books, 2000.

8. Interview with Alexander Rose, San Francisco, CA, June, 2017.

# CHAPTER 7

# Conclusions and Trajectories: Future Cities that are Blue and Biophilic

**Abstract** Blue biophilic cities represent a compelling new urban vision for the future, one appropriate both for the opportunities available and for the challenges faced by coastal cities. The marine nature all around can be the source of immense benefit for urban residents in the form of wonder, physical and mental health, purpose and meaning. At the same time, blue biophilic cities can exert leadership in ocean conservation and serve as a pioneer in shaping new and more sustainable relationships with the marine world (e.g. by re-thinking what they harvest from and grow in the ocean). It is hoped that the stories told, and the examples and ideas presented, in this book are just the beginning, and that blue biophilic cities will continue to innovate and continue to imagine and explore new connections to the marine environment.

The main message of this book is that we need blue nature in cities, and in turn oceans need positive leadership and planning from cities. The book builds on the basic insight of biophilia—that we are innately drawn to nature and living systems, and contact with nature is important and essential to healthy and meaningful human lives. The vision of biophilic cities—cities that are profoundly natureful, that seek to foster connections to the natural world, and where residents are curious about and attentive to the nature around them—is gaining traction in the USA and around the

© The Author(s) 2018

T. Beatley, *Blue Biophilic Cities*, Cities and the Global Politics of the Environment, https://doi.org/10.1007/978-3-319-67955-6_7

world, and a new global Biophilic Cities Network has been formed to help to advance this global movement. Our notion of biophilic cities can't just rely on a terrestrial view of the world, not on a blue planet: it must increasingly understand them as cities connected to and embedded in a larger marine fabric. All cities, of course, even interior land-locked cities, are affected by and in turn affect the ocean world, but it is the coastal and port cities where the potential to cultivate a truly blue biophilic urbanism is most evident. And we have some good examples of cities that are doing this.

These new models of blue and biophilic urbanism, and our new global Biophilic Cities Network, are emerging at just the right time. Globally, we are engaged in a collective discourse about cities and the environment, and there is a growing recognition that current directions must change. In 2015, many countries adopted the *2030 Agenda for Sustainable Development*, and the 17 Sustainable Development Goals (building on the earlier Millennium Goals). As Mentioned earlier, two goals stand out in their connection to blue biophilic cities: Goal 11 (Sustainable Cities and Communities) and Goal 14 (Ocean Conservation). There are more specific targets identified for each of these goals, including, for instance, ending overfishing by 2020 (a Goal 14 target), and, by 2020, ensuring "universal access to safe, inclusive and accessible, green and public spaces, in particular for women and children, older persons and persons with disabilities."[1] Bluespaces and blue nature in cities will be one important avenue for achieving this target, and more generally the vision of cities I am advancing here will help to address many (most) of the Sustainable Development Goals.

## WE NEED NATURE AND WILDNESS IN OUR LIVES MORE THAN EVER AND THE OCEAN CAN BE A MAJOR SOURCE

We want and need connection to blue nature, and, as this book has argued throughout, we gain much from this connection—we are happier, healthier, able to live more meaningful lives, and lives of (daily) wonder and awe. Equally true, cities can and must be advocates for ocean conservation, and for the diverse marine life that relies on healthy oceans. Cities must be leaders in advancing a planning and design agenda that acknowledges a deeper biophilic sensibility that recognizes the inherent value of marine life.

The marine world represents at least a significant part of the solution to our current disconnect from nature. As the creative efforts reported here show, from Pier Into the Night to efforts of ocean educators such as the Biscayne Nature Center, to the Billion Oyster Program, we can indeed re-connect to nature and living systems, and in and near cities we can feed and support our innate biophilic sensibilities. They are there waiting to find expression and waiting to be cultivated. Much of this can and must happen by reforming our schools so that they, in the spirit of the New York Harbor School, acknowledge and embed learning in and through the local marine environment. Marine nature is often all around us, and one of the clearest forms of nearby nature and wildness that we must take advantage of. As Fisher told me, "we've got to get city kids comfortable with being around the water. That is an absolute responsibility we have" (Fig. 7.1).

Over the course of the preceding chapters, certain key themes have emerged. In this final chapter I summarize these and offer some speculation about future directions they might move in or evolve towards.

## We Need to Creatively Balance the Danger and the Delight

Coastal cities from New York to Shanghai will experience serious challenges in adapting to accelerating sea level rise and coastal flooding. Many cities are now discovering new and creative ways to both adapt to climate change and sea level rise and enhance connections to nature. The coastal edge is being re-defined in many blue cities in some promising new ways, as reflected in new projects such as the Dryline. In some places (perhaps many), significant investments in more traditional shore armoring strategies will occur and increase (seawalls, revetments, jetties, floodgates), but in many places there is a shift to a model of a more dynamic shoreline, one that is designed to permit or accommodate some degree of flooding. And cities such as New York understand the need for some degree of retreat, both horizontal (floodable edges that can be converted from roads, parking lots and hard surfaces to wetlands and permeable spaces) and vertical (as we see in the elevation of new structures such as the American Copper Building in New York). There is the potential that in blue cities, as in the case of the Dryline, adapting to sea level rise will also present opportunities to grow more urban nature and to connect residents to the marine realm around them.

**Fig. 7.1**  Students of the New York Harbor School help to grow and monitor oysters as part of the Billion Oyster Project. Image credit: New York Harbor School

Also promising are the many ways in which cities are designing shorelines in more natureful ways, including floating wetlands, living shorelines and breakwaters, and ecological seawalls that are designed to accommodate greater marine biodiversity and to mimic natural shorelines and rocky edges. Even more creative work in these areas will likely happen in the years ahead. Singapore's efforts at research and development regarding different shoreline structures and their relative effectiveness at accommodating marine organisms is a good sign. The country has been a leader in testing and applying different designs to incorporate nature into high-rise buildings (including a green wall long-term monitoring site at the HortPark).

I have sought more generally in this book to temper the sense of impending doom that tends to take over in discussions about sea level rise. We are drawn to water, and the evidence is growing that proximity to water, physical and visual access to water, and being immersed in water have therapeutic and healing powers. Water is one of those extremely important (and ancient) biophilic qualities and place attributes that humans want and need, something that blue biophilic cities understand.

## We Need to Appreciate the Ocean as Medicine

In these ways we might begin to better appreciate the ocean medicine that surrounds coastal cities and to design and plan accordingly. Blue biophilic cities recognize the marine context as an incredible asset to take advantage of in enhancing physical and mental health. Cities such as New York are re-discovering the power of water and returning to earlier mottos of "city of water" as a way of self-describing or self-defining.

In cities such as Seattle there are new and bold efforts to re-design the urban fabric to provide new physical connections to water. The Seattle Waterfront Plan is an ambitious scheme to re-connect the city to the water, with new park space, a continuous promenade, re-built piers, places to launch kayaks, and new street level connections to surrounding neighborhoods. There are new human–water connections and also new investment for marine nature, including the re-design of a main seawall to provide a habitat for salmon, habitat benches and light-penetrating sidewalk panels, and a set of unique textured surfaces that create spaces for marine invertebrates (Fig. 7.2).

It is interesting to realize that as a species we are just in the early stages of developing an ethics and ethos of appreciating the complexity, beauty

**Fig. 7.2** New York City has opened up many new parks along its waterfront in recent years, including this one along the Hudson River, adjacent to one of the city's most popular bike routes. Image credit: Tim Beatley

and wonder of the marine world, and caring for its condition. I am reminded of the remarkable work of early naturalists such as Philip Henry Gosse, who coined the word "aquarium" and helped to design and perfect the first versions of them, including the first public aquarium at Regent's Park in London, opening in 1853.[2] His 1854 book *Aquarium* helped to ignite the early Victorian craze for aquariums, and his later book *Oceans*, and other prolific writings and work, did much to ignite a popular interest in ocean life.[3] Referred to now as the David Attenborough of his time, he was an especially eloquent voice for the majesty of the natural world. Before underwater photography, his wonderful watercolors of marine organisms conveyed their beauty and otherworldliness. An early marine scientist, he also found many ways to share his love of ocean life, including low-tide nature walks along the Devon coast, in a way presaging some of the ways that today we educate about and connect with oceans.

Gosse's work was just 160 years ago, we should remember, so we have not long had the view of oceans as places to be fascinated by, and to protect and conserve. Our usual approach to oceans was more as a repository for demons and dragons and as intrinsically dangerous, sinister places. Squarely in the urban age today, cities must extend and expand this view of the wonder below. A terrific biography of Gosse by Ann Thwaite is aptly titled *Glimpses of the Wonderful*, a famous Gosse quote, but also a positive goal and measure of the success of every blue biophilic city.[4]

## We Must Continue to Re-wild Our Blue Cities

Increasingly we are understanding our cities as places of nature, as we see coyotes, foxes and other mammals that sometimes present co-existence challenges, but that infuse elements of beauty and wildness into contemporary cities. Initiatives such as Wildlife Watch in Chicago seek to better understand the extent of animal life living in and adapting to urban environments. In that example some 120 camera traps are deployed along transects through the city in one of the first attempts to comprehensively study urban wildlife. Similar efforts might be taken to better understand and study the marine life near cities which is even more invisible. Camera traps are yielding important insights into urban wildlife adaptation but just as importantly they are helping to educate about the animals that Chicagoans share their city with. It is an interesting question how similar efforts might be focused on aquatic and marine environments near cities (deploying underwater cameras and hydrophones perhaps?).

Arguably the marine habitats near to cities represent a level of wildness unmatched by anything on land. Here we have habitats that while negatively impacted by pollution, for example, have not been occupied by humans in the same way as terrestrial lands, and still retain a remarkable level of biodiversity. These marine environments serve as home to a remarkable array of organisms. The extent of the wildness becomes apparent when cities such as Singapore attempt to study it. One of the few cities to have conducted a Comprehensive Marine Biodiversity Survey (every blue biophilic city should do this), this five-year study collected some 30,000 specimens, utilizing scientists and volunteers, and uncovered some 14 marine species that were entirely new to science, including a "lipstick" sea anemone, and an orange-clawed mangrove crab.

Our sense of wonder about the marine world is enhanced immeasurably by how much we don't know and what we are still learning. Only since 1968 have Katy and Roger Payne discovered the humpback whale songs, it might be remembered, with new insights unfolding about how these cetaceans modify and re-mix these songs over time.[5] It was only half a century ago, yet the discovery of these songs changed our view of whales and the songs themselves have become iconic. We are also learning more, almost daily it seems, about the extent of the unusual sea life at greater depths. An ongoing Australian expedition using low-tech collection methods is finding a variety of unbelievable and fascinating animals, including coffinfish, sea spiders and an abundance of echinoderms on the seafloor affectionately known as "sea pigs."[6] Again, it would be hard to invent these creatures for a sci-fi Hollywood film. Even marine species we think we know well are proving to be powerfully different. Philosopher Peter Godfrey-Smith's book *Other Minds* explores and seeks to explain the evolution of the brains of cephalopods, so distant from the evolutionary lineage of the human brain. As he provocatively declares, an octopus "is the closest we will come to meeting an intelligent alien".[7]

Murray Fisher spoke passionately in our interview about the promise of re-wilding New York Harbor. As a practical matter, this could happen in many different ways. Coastal cities can and should find ways to support the establishment of marine parks and protected areas. A biophilic city is one that seeks to grow and expand the nature in and around cities, but also understands a broader duty to care for and support (indeed assume leadership for) larger global conservation efforts. With respect to oceans, this means significant expansion of protected marine areas. As discussed earlier, the goal of setting aside 30 % of the planet's ocean surface in protected marine areas is a good one, though 50 % would be bolder and more ambitious.[8] Support for a connected network of such areas by blue biophilic cities is essential. There is a special role to be played by nearby marine parks, even quite small ones. A number of examples can be found in the coastal suburbs around Sydney (e.g. Gordon's Bay Aquatic Reserve). The Taputeranga Marine Reserve in Wellington and the Island Bay Marine Education Centre are also positive examples. These close-by marine parks are the places where urban residents can cut their teeth on marine biology, can be enticed to dip their biophilic toes, to experience something of the wonder of the marine world without traveling a great distance by sea or plane (Fig. 7.3).

## CITIES MUST BECOME A POTENT FORCE ON BEHALF OF THE OCEANS

In July, 2017, the Australian federal government announced significant new reductions in the no-take zones of its network of protected marine areas. Its ambitious system for marine protection gained deserved acclaim when it was unveiled in 2012, but from the start there were pressures to weaken it.[9] The Australian story highlights the pitfalls of failure to fully mobilize a constituency that would object to such proposals. Cities, and urban populations, can and should serve in this capacity, and especially so given their proximity to where these marine conservation zones are located.

The main point here is that we need to find ways to cultivate and activate urban populations and cities on behalf of ocean conservation. This means both local and global action, but suggests the value of local efforts at blue re-wilding as a point of connection and an entree into re-wilding that needs to occur beyond harbors and sounds and near-to-shore waters.

## WE NEED TO SECURE OUR DAILY DOSES OF BLUE NATURE

A challenge is to find ways to cultivate daily experiences of marine nature in blue cities. We are often articulating the experience of nature in biophilic cities in terms of an urban nature diet, and posing the question of what constitutes a healthy urban nature diet. We find it helpful to visualize this in terms of a pyramid—the nature pyramid (a parallel to the food pyramid intended to inform healthy food choices). At the top of the pyramid are more immersive experiences of nature obtained by visiting a faraway

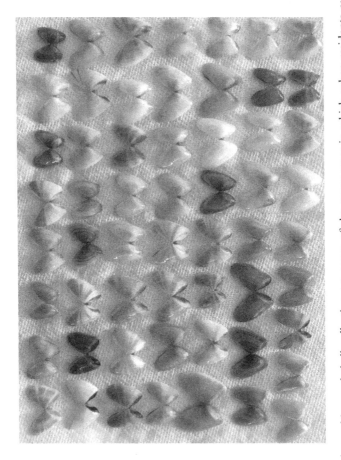

**Fig. 7.3** Beach combing and shell collecting are some of the many ways in which urban residents can enjoy nearby marine environments. One of my family's favorite pastimes is collecting and creatively displaying colorful coquina clam shells. Image credit: photo by Tim Beatley; credit for the idea of displaying coquina's in this way, and by creatively framing them, goes to Anneke Bastiaan

park or natural landscape. Building a nature diet around these types of experience is impractical and costly (and costly from the perspective of the carbon footprint associated with visiting such places). On the other hand, we must move down the pyramid and look for ways to experience everyday nature near to where we live and work—these should make up the bulk of our urban nature diet. To date, we have mostly envisioned this pyramid in terms of terrestrial habitats and settings—trees, birds, urban parks and gardens, and a host of biophilic design strategies such as living walls and eco-roofs. But the pyramid for a blue biophilic city should contain and reflect marine opportunities, whether that might be an hourly scan of a harborscape or seascape, a daily swim or coastal walk, a weekend scuba dive or snorkel, or (moving up the pyramid) a more distant marine-oriented holiday. Figure 7.4 presents at least one possible version of the blue nature pyramid, and every blue biophilic city should work to include in its nature diet a healthy dose of the blue.

Blue biophilic cities must continue to explore the idea of integrated land–sea parks. The recent examples of the Hudson River and Brooklyn Bridge Parks in New York City, which place contact with and connection to water at their core, are also exemplars in the direction of this positive trend.

Part of the agenda of re-wilding is about continuing the investments made in many cities in cleaner water. Improving water quality in cities such as New York and Boston pays dividends in many ways, including in the biodiversity and marine life supported by the ecosystems. The return of humpback whales, and of the menhaden fish they follow, to New York Harbor shows the fruits of these investments.

## We Must Work to Make Blue Nature More Visible

As argued at several points in this book, the invisibility and emotional remoteness of the sea make it difficult to imagine and care for the oceans in cities. Much of the marine environment is physically remote and far away, to be sure, but much of it is remarkably proximate.

We have reviewed a number of innovative strategies for making blue nature and blue wildness visible. What can be done with a simple low cost video camera, a few underwater lights and a couple of volunteer divers can be seen in the story of the Pier Into the Night events in Gig Harbor, Washington. Efforts to guide school-age kids into the water, and to invite them to find and explore the marine nature under their very feet, with the help of skilled naturalists, is another approach.

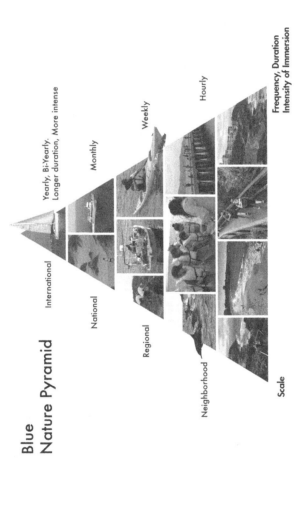

**Fig. 7.4** The blue urban nature pyramid provides a way to begin to understand and visualize the minimum amounts of daily marine nature we need and might have available in a blue biophilic city. Image credit: original concept by Tanya Denckla-Cobb; image by Tim Beatley

And there are effective ways we can continue to make even the faraway marine world visible and present. The wonderful underwater photography of people such as Brian Skerry and Anne Doubilet has a power to move and connect.

Finding ways to take full advantage of those "glimpses of the wonderful" that occur at certain times in many cities—low tide being one of them—is another strategy, and schemes such as the Seattle Aquarium Beach Naturalist program is one such example. Here, trained volunteers help strolling visitors to several of the city's shoreline parks understand and appreciate the marine organisms they see there.

Equally true, we must design the next generation of urban waterfront and harborfront buildings and neighborhoods in ways that offer physical connections to the water, and ideally opportunities to see and understand the marine life nearby.

As Jane Lubchenco has said,

> Most of us live in cities and because of that it will really be incumbent on us to take the lead in being better stewards of our planet, stewards of the land, and stewards of the ocean. Even though most of the ocean is far away from where any of us live, we can still play a very active role in championing, protecting important places, protecting healthy ocean ecosystems, protecting iconic species. And as individuals we can work together with like-minded people.

As the examples in this book show, there are now wonderful programs that engage the public directly in the restoration of marine environments. The BOP is one that gets kids and schools involved in restoring New York Harbor.

And while we bemoan the time kids and adults alike spend online and on social media, we are increasingly finding new and creative ways to uses this technology to connect people to marine environments and organisms. Through apps such as Shark Tracker, which follows in real time the movements of great whites such as Mary Lee, fear may give way to familiarity and kinship. Such technology moreover may help make more traditional ways of connecting with marine life—such as whale watching—easier. An Australian app, Wild About Whales, for instance, helps residents of Sydney track sightings of whales in real time, helping to watch for and see these animals.[10]

A biophilic marine or ocean city finds other ways to foster a sense of connection. In biophilic design, much attention is paid also to the shapes and forms of nature, and to images of nature, not just living nature. The Australia Port City of Fremantle, for instance, has invested in a variety of different forms of public art, but much of it takes local nature as its subject, and many of the nods to nature are marine. Entering the city on one of the main roads from Perth, one is confronted by a large mural of octopus, for instance, and there are many other places in the city where art is embedded in sidewalks and in murals. Biophilic art can find expression and support in many ways, and in Fremantle the city implements a One Percent For Art requirement—1 % of the budget of all development projects, private as well as public, must be set aside to support public art. These investments enhance quality of life and our connections to the natural world, though the evidence of the therapeutic and healing power natural images is much less definitive than for exposure to living nature (Fig. 7.5).

## We Must Adjust Our Mental Maps of Cities

A big part of the agenda of blue biophilic cities is the cultivation of a new imagination about cities, one that sees them encompassing and extending to include the marine world around them. It is in large part a matter of re-drawing our mental maps of cities, which until now have tended only to see empty or blank spaces beyond the shore's edge. And too often we literally draw our maps—sometimes plan diagrams, or official land use planning and policy maps—as though there is nothing there. There are certainly limits to the jurisdictional planning and control that cities can exert, but even when there is no clear power to plan beyond mean high tide there is a broader purpose served—to extend our consideration of and care for this marine world, to reinforce visually that these are natureful, biodiverse, wondrous places that need to be protected and conserved. I have frequently com-mented on the official planning maps of cities such as San Francisco, that show well the parks and greenspaces on land but seem to indicate just an empty void (or a profound lack of interest in) beyond the terrestrial edge.

Ideally blue biophilic cities are able to extend even further their spatial sense of connection. Along the East Coast of the USA, for instance, lies a network of deep canyons, many specifically named after the nearest land-based city. There is a Norfolk Canyon, named after the Virginia city, for example. Through a blue biophilic frame, these are wonderful areas of the sea that cities (like Norfolk) could adopt or embrace in some way and be proud of. Recent underwater expeditions have shown a remarkable marine life there, and although few Norfolk citizens are likely to visit this underwater

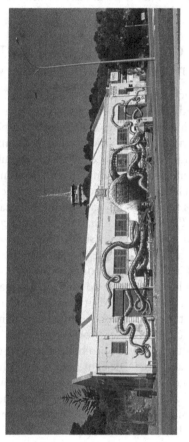

**Fig. 7.5**  Octopus mural, Fremantle, Western Australia. Image credit: Paul Weaver

marine habitat, their lives may be enriched by knowing about it. And the life of that underwater canyon can certainly use the conservation friend. One idea that extends our spatial frame, and that was proposed in an earlier book, is that of an ocean sister city, and perhaps the Norfolk Canyon could be so adopted by the City of Norfolk. It should also contribute to a city's pride of place.[11]

## WE MUST CONTINUE TO DEVELOP THE VISION AND PRACTICE OF BLUE AND BIOPHILIC CITIES

We are at a unique juncture in the history of the blue planet, one in which population growth and levels of consumption have soared, and more and more of us are living in cities. We have also reached a point where many, perhaps most of us, living in cities have become profoundly disconnected from nature and natural systems. The concept of biophilia argues that we are drawn to nature and living systems, that there is an innate pull that nature exerts on us. The emerging research in psychology, medicine and public health confirms the many physical and mental health benefits delivered by nature, and I have argued that these connections can provide meaning, purpose and a deeper dimension to life. The vision of biophilic cities, cities that put nature at the core of their design and planning, that are natureful and seek to foster deep connections with the natural world, is taking hold. The blue dimension to this agenda has been slower to materialize and harder to envision. But that can and must change, and indeed is changing, as the case studies described here demonstrate. We will need to continue to expand these efforts, to further flesh out what it means to be a blue biophilic city, and to build a set of programs and initiatives, coastal projects and urban designs, and many other exemplars that show how cities can connect with blue nature. This will be the task of many—city planners, architects and engineers, marine conservationists and, of course, the residents of the cities where the urban blue can do so much to frame and enhance future urban life.

## NOTES

1. See "Goal 11 Targets," found at: http://www.un.org/sustainabledevelopment/cities/
2. For an excellent review of that history, see Tim Wijgerde, "Victorian Pioneers of the Marine Aquarium," found at: http://www.advancedaquarist.com/2016/2/aafeature2

3. Philip Henry Goss, *The Aquarium: An Unveiling of the Wonders of the Deep Sea*, London: J. Van Voorst, 1856.
4. Ann Thwaite, *Glimpses of the Wonderful: The Life of Philip Henry Goss*, Faber and Faber, 2002.
5. Ed Yong, "Humpback Whales Remix Their Old Songs."
6. "Australia's Most Bizarre Creatures Uncovered in Deep Sea Expedition," found at: https://particle.scitech.org.au/earth/australias-bizarre-creatures-uncovered-deep-sea-expedition/
7. Peter Godfrey-Smith, *Other Minds: The Octopus, the Sea, and the Deep Origins of Consciousness*, Farrar, Straus and Giroux, 2016.
8. E.O. Wilson, *Half-Earth: Our Planet's Fight for Life*, Norton, 2017.
9. Michael Slezak, "Australia's Marine Parks Face Cuts to Protected Areas," *The Guardian*, July 21, 2017, found at: https://www.theguardian.com/environment/2017/jul/21/turnbull-government-plans-further-cuts-to-fishing-protection-zones?CMP=share_btn_tw
10. https://www.wildaboutwhales.com.au/app
11. See the discussion of the idea of ocean sister cities in Beatley, *Blue Urbanism*, Island Press, 2014, pp. 152–153.

# BIBLIOGRAPHY

Aspinall, Peter, Mavros, Panagiotis, Coyne, Richard, and Roe, Jenny. 2015. The Urban Brain: Analyzing Outdoor Physical Activity with Mobile EEG. *Journal of British Sports Medicine*, March, Feb, 49(4):272–276.

Barton, Jo, and Jules Pretty. 2010. What Is the Best Dose of Nature and Green Exercise for Improving Mental Health? A Multi-Study Analysis. *Environmental Science and Technology* 44: 3947–3955.

Beatley, Timothy. 2009. *Planning for Coastal Resilience*. Washington, DC: Island Press.

———. 2011. *Biophilic Cities: Integrating Urban Design and Nature*. Washington, DC: Island Press.

———. 2014a. Launching the Global Biophilic Cities Network. http://www.thenatureofcities.com/2013/12/04/launching-the-global-biophilic-cities-network/

———. 2014b. *Blue Urbanism: Connecting Cities and Oceans*. Washington, DC: Island Press.

———. 2017. *Handbook of Biophilic City Planning and Design*. Washington, DC: Island Press.

Beatley, Timothy and Peter Newman. 2013. Biophilic Cities Are Sustainable, Resilient Cities. *Sustainability*, June. http://www.mdpi.com/2071-1050/5/8/3328

Bekoff, Marc. 2014. *Rewilding Our Hearts: Building Pathways of Compassion and Coexistence*. Novato, CA: New World Library.

Blaustein, Richard. 2014. Urban Biodiversity Gains New Converts: Cities Around the World Are Conserving Species and Restoring Habitat. *BioScience*. http://bioscience.oxfordjournals.org/content/63/2/72.full

© The Author(s) 2018                                                      133
T. Beatley, *Blue Biophilic Cities*, Cities and the Global Politics of the Environment, https://doi.org/10.1007/978-3-319-67955-6

Bratman, Gregory N., J. Paul Hamilton, and Gretchen Daily. 2012. The Impacts of Nature Experience on Human Cognitive Function and Mental Health. *Annals of the New York Academy of Sciences* 1249: 118–136.

Browning, William. 2014. The 14 Patterns of Biophilic Design. http://www.terrapinbrightgreen.com/report/14-patterns/

Carson, Rachel. 1951. *The Sea Around Us.* Oxford: Oxford University Press.

Cooper Marcus, Clare, and Naomi Sachs. 2013. *Therapeutic Landscapes: An Evidence-Based Approach to Designing Healing Gardens and Restorative Outdoor Spaces.* New York: Wiley Press.

Donovan, Geoffrey, and David T. Butry. 2010. Trees in the City: Valuing Street Trees in Portland, Oregon. *Landscape and Urban Planning* 94: 77–83.

Dooling, Sarah. 2009. Ecological Gentrification: A Research Agenda Exploring Justice in the City. *International Journal of Urban and Regional Research* 33 (3): 621–639.

Douglas, Marjory Stoneman. 1947. *The Everglades: River of Grass.* New York: Rinehold and Company.

Earle, Sylvia. 2010. *The World Is Blue: How Our Fate and the Ocean's Are One.* Washington, DC: National Geographic.

Gosse, Philip Henry. 1856a. *The Aquarium: An Unveiling of the Wonders of the Deep Sea.* London: J. Van Voorst.

———. 1856b. *The Ocean.* London: Society for Promoting Christian Knowledge.

Hartig, Terry, Marlis Mang, and Gary Evans. 1991. Restoration Effects of Natural Environment Experiences. *Environment and Behavior* 23: 3.

Kellert, Stephen. 2014. *Birthright: People and Nature in the Modern World.* New Haven, CT: Yale University Press.

Kellert, Stephen R., Judith Heerwagen, and Martin Mador. 2008. *Biophilic Design: The Theory, Science and Practice of Bringing Buildings to Life.* Hoboken, NJ: Wiley.

Kellert, Stephen, and E.O. Wilson. 1995. *The Biophilia Hypothesis.* Washington, DC: Island Press.

Louv, Richard. 2008. *Last Child in the Woods: Saving Our Children From Nature-Deficit Disorder.* Chapel Hill, NC: Algonquin Books.

———. 2012. *The Nature Principle: Reconnecting with Life in a Virtual Age.* Chapel Hill, NC: Algonquin Books.

Nichols, Wallace J. 2014. *Blue Mind.* New York: Little, Brown and Company.

Orff, Kate. 2016. *Toward an Urban Ecology.* New York: Monacelli Press.

Patel, Neel V. 2014. Migrating to the City: How Researchers Are Beginning to Think Differently About Urban Biodiversity. *Science Line.* http://scienceline.org/2014/06/migrating-to-the-city/

Pauly, Daniel, and Dirk Zeller, eds. 2016. *Global Atlas of Marine Fisheries: A Critical Appraisal of Catches and Ecosystem Impacts.* Washington, DC: Island Press.

Selhub, Eva, and Alan Logan. 2012. *Your Brain on Nature: The Science of Nature's Influence on Your Health, Happiness and Vitality.* New York: Wiley Press.

Stoner, Tom, and Carolyn Rapp. 2008. *Open Spaces, Sacred Places.* Baltimore, MD: TKF Foundation.

Terrapin Bright Green, The Economics of Biophilia. http://www.terrapinbright-green.com/report/economics-of-biophilia/

Thomas, Sue. 2013. *Technobiophilia: Nature and Cyberspace.* New York: Bloomsbury Academic.

Tidball, Keith. 2014. Urgent Biophilia: Human-Nature Interactions in Red Zone Recovery and Resilience. In *Greening in the Red Zone: Disaster, Resilience and Community Greening*, ed. K.G. Tidball and M.E. Krasny, 50. The Netherlands: Springer.

Tova Bailey, Elizabeth. 2010. *The Sound of a Wild Snail Eating.* Chapel Hill, NC: Algonquin Books.

Van der Wal, Ariane, Hannah Schade, Lydia Krabbendam, and Mark van Vugt. 2013. Do Natural Landscapes Reduce Future Discounting in Humans? *Proceedings of the Royal Society B* 280. https://doi.org/10.1098/rspb.2013.2295.

Weinstein, Netta, Andrew K. Przybylski, and Richard M. Ryan. 2009. Can Nature Make Us More Caring? Effects of Immersion in Nature on Intrinsic Aspirations and Generosity. *Personality and Social Psychology Bulletin* 35: 1315–1329.

Wilson, Edward O. 1984. *Biophilia.* Cambridge, MA: Harvard University Press.

———. 2007. *The Creation: An Appeal to Save Life on Earth.* New York: Norton and Company.

Wolch, Jennifer R., Jason Byrne, and Joshua P. Newell. 2014. Urban Green Space, Public Health, and Environmental Justice: The Challenge of Making Cities 'Just Green Enough'. *Landscape and Urban Planning* 125: 234–244.

# Index[1]

[1]Note: Page numbers followed by 'n' refer to notes.

© The Author(s) 2018                                                     137
T. Beatley, *Blue Biophilic Cities*, Cities and the Global Politics of the Environment, https://doi.org/10.1007/978-3-319-67955-6

Printed in the United States
By Bookmasters